Exercise 1   Mineral Key

DO NOT BREAK THE MINERALS—THEY ARE ALREADY BROKEN!!!

Moh's Hardness Scale:

1. talc

2. gypsum

3. calcite

4. fluorite

5. **apatite**

6. orthoclase feldspar

7. quartz

8. topaz

9. **corundum**

10. diamond

Varieties of Quartz:

11. rose quartz

12. milky quartz

13. smoky quartz

14. amethyst

15. **quartz**

Varieties of Feldspars:

16. orthoclase feldspar

17.   "   "

18. plagioclase feldspar

19.   "   "

Other Minerals:

20. kaolinite

21. muscovite

Ferromagnesians:

22. biotite

23. hornblende

24. augite

25. olivine

Other Minerals:

26. pyrite

27. chalcopyrite

28. galena

29. sphalerite

30. barite

31. **garnet**

32. halite

33. hematite

34. magnetite

Other minerals you will see on practice quizzes and thus expected to know are:

35. copper

36. sulfur

37. graphite

38. malachite

is a crystal

CONTENTS

Kendall Hunt
publishing company

www.kendallhunt.com
Send all inquiries to:
4050 Westmark Drive
Dubuque, IA 52004-1840

EARTH SCIENCE

Revised Printing

LAB MANUAL

Robert Mims

Richland College

Dallas County Community College District

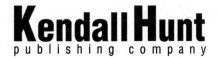
Kendall Hunt
publishing company

Report Sheet   Name:

Report Sheet   Name:

Mineral Practice Quiz 1   Lab:

Name each mineral:

1 _____   24 _____

2 _____   25 _____

3 _____   26 _____

4 _____   27 _____

5 _____   28 _____

6 _____   29 _____

7 _____   30 _____

8 _____

9 _____

10 _____

11 _____

12 _____

13 _____

14 _____

15 _____

16 _____

17 _____

18 _____

19 _____

20 _____

21 _____

22 _____

23 _____

Exercise 2   Report Sheet   Name: _____

Mineral Practice Quiz 2   Lab: _____

State the physical property:

1 _____ (luster)   24 _____

2 _____ (luster)   25 _____

3 _____ (hardness)   26 _____

4 _____ (hardness)   27 _____

5 _____ (cleavage type)   28 _____

6 _____ (cleavage type)   29 _____

7 _____ (# cleavages)   30 _____

8 _____ (# cleavages)

Name each mineral:

9 _____

10 _____

11 _____

12 _____

13 _____

14 _____

15 _____

16 _____

17 _____

18 _____

19 _____

20 _____

21 _____

22 _____

23 _____

Exercise 3   Report Sheet   Name:

Mineral Practice Quiz 3   Lab:

State the physical property:

1 _____ (luster)   24 _____

2 _____ (luster)   25 _____

3 _____ (hardness)   26 _____

4 _____ (hardness)   27 _____

5 _____ (cleavage type)   28 _____

6 _____ (cleavage type)   29 _____

7 _____ (# cleavages)   30 _____

8 _____ (# cleavages)
Name each mineral:

9 _____

10 _____

11 _____

12 _____

13 _____

14 _____

15 _____

16 _____

17 _____

18 _____

19 _____

20 _____

21 _____

22 _____

23 _____

Exercise 4   Report Sheet   Name:

Mineral Practice Quiz 4   Lab:

State the physical property:

1 _____ (luster)   24 _____

2 _____ (luster)   25 _____

3 _____ (hardness)   26 _____

4 _____ (hardness)   27 _____

5 _____ (cleavage type)   28 _____

6 _____ (cleavage type)   29 _____

7 _____ (# cleavages)   30 _____

8 _____ (# cleavages)

Name each mineral:

9 _____

10 _____

11 _____

12 _____

13 _____

14 _____

15 _____

16 _____

17 _____

18 _____

19 _____

20 _____

21 _____

22 _____

23 _____

Exercise 5   Instructions

Density

Use a pencil.

If you do not have a calculator then use the computer:
Click Start, Programs, Accessories, then Calculator.
Perform the procedures in random order because there is not enough equipment for everyone
to do the same procedure simultaneously.

Procedure 1 (Density of Wooden Block):

1. Do not share wooden blocks. Select a wooden block and measure and record its dimensions in centimeters using the plastic ruler. The measurements should all have two decimal places.

2. Calculate the volume of the wooden block in **Data Table 1.** Write the formula first. Beneath it substitute the measurements complete with units in lieu of the symbols in the formula. Beneath it write the answer <u>complete with units</u> and circle them:

    Example: V = LWH
                                            V = (2.55 **CM**) (4.15 **CM**) (3.00 **CM**)
                                            V = 31.75 **CM**^{3}

3. Calculate the density of the wooden block (using the formula in your classnotes) in **Data Table 2.** Be sure to write the units of each measurement on each line.

4. Match your calculated density with one of those which is reasonably close in **Data Table 2** and record the type of wood. It is unlikely that your calculated density will be a perfect match.

	Densities of Common Woods:		
Wood:	Density:	Wood:	Density:
1. balsa	0.20 g/cm^3	7. walnut	0.76 g/cm^3
2. redwood	0.41 g/cm^3	8. mahogany	0.48 g/cm^3
3. cypress	0.27 g/cm^3	9. cherry	0.75 g/cm^3
4. fir	0.54 g/cm^3	10. white pine	0.40 g/cm^3
5. white oak	0.78 g/cm^3	11. yellow pine	0.48 g/cm^3
6. magnolia	0.60 g/cm^3	12. maple	0.82 g/cm^3

Procedure: Density of Water:

1. Select a **small** plastic (10 ml) graduated cylinder. Dry it if necessary.

2. Determine the mass of the cylinder on the electronic balance and record the value in **Data Table 3.**

3. Add 10 ml of water to the plastic cylinder. Read from the bottom of the curved water surface (meniscus) not the top. Record the volume in **Data Table 3.** Then determine the mass of the <u>cylinder with water</u> but be sure no droplets are clinging to the walls of the cylinder.

4. Calculate the density of water in **Data Table 3.**

Procedure: Density of Quartz:

1. Take a quartz crystal and determine its mass and record in **Data Table 4.**

2. Take the **large** plastic (50 ml) graduated cylinder and fill it about half full of water and record the volume in **Data Table 4.**

3. Tilt the cylinder to a low angle and slowly slide the quartz crystal into the water. Record the new water level.

4. Take the difference between water levels and record as the volume of the quartz crystal.

5. Calculate the density of quartz in **Data Table 4.**

Procedure: Density of **Narrow** Aluminum Cylinder

1. Take a **narrow** aluminum cylinder and complete **Data Table 5** using the same (submersion) technique that is used for quartz in the above procedure.

Procedure: Volume of **Wide** Aluminum Cylinder

1. Take a **wide** aluminum cylinder and complete **Data Table 6.** To determine the radius it is easier to measure the diameter and divide by 2. To calculate the volume of a cylinder the formula is:

    $V = \Pi R^2 H$ where $\Pi = 3.14$, R is radius and H is height.

Procedure: Density of **Wide** Aluminum Cylinder

1. Calculate the density of the **wide** aluminum cylinder in **Data Table 7.**

2. Please return everything.

_____

Report Sheet   Name:

_____

Density   Lab:

_____

| Data Table 1 (Wood Volume) | Data Table 2 (Wood Density) |

_____is wood block number   _____g = mass of wood block

WRITE TWO DECIMAL PLACES:
Length = _____cm
Width   = _____cm
Height = _____cm

_____

Calculate **wood** block volume: Calculate **wood** block density:

(WRITE FORMULA:)   (WRITE FORMULA:)

V =   D =

(SHOW CALCULATIONS / UNITS / 2 DECIMAL PLACES:)   (SHOW CALCULATIONS / UNITS / 2 DECIMAL PLACES:)

V =   D =

(WRITE ANSWER / UNITS / 2 DECIMAL PLACES:)   (WRITE ANSWER / UNITS / 2 DECIMAL PLACES:)

V =   D =

                                                     _____is the type of wood

_____

Data Table 3 (Water Density)   Data Table 4 (Quartz Density)

_____g = mass of cylinder   _____g = mass of quartz
                    (empty)
_____g = mass of cylinder   _____ml = water level
                    with water   (without quartz)
                                                     _____ml = water level
_____g = mass of water   with quartz
                                                     _____ml = difference in
_____ml = volume of water   water levels
                                                     _____cm^3 = volume of quartz

_____

Calculate density of water: Calculate density of quartz:

(WRITE FORMULA:)   (WRITE FORMULA:)

D =   D =

(SHOW CALCULATIONS / UNITS / 2 DECIMAL PLACES:)   (SHOW CALCULATIONS / UNITS / 2 DECIMAL PLACES:)

D =   D =

(WRITE ANSWER / UNITS / 2 DECIMAL PLACES:)   (WRITE ANSWER / UNITS / 2 DECIMAL PLACES:)

D =   D =

14

Data Table 5 (Aluminum Density)

Narrow aluminum cylinder density:

_____g = mass of **narrow** aluminum cylinder

_____ml = water level (without aluminum)

_____ml = water level with aluminum

_____ml = difference in water levels

_____cm^3 = volume of **narrow** aluminum cylinder

Calculate density of **narrow** aluminum cylinder:

(WRITE FORMULA:)

D =

(SHOW CALCULATIONS / UNITS / 2 DECIMAL PLACES:)

D =

(WRITE ANSWER / UNITS / 2 DECIMAL PLACES:)

D =

Data Table 6 (Aluminum Volume)

Wide aluminum cylinder volume:

_____g = mass of **wide** aluminum cylinder

WRITE TWO DECIMAL PLACES:

Height = _____cm

Diameter = _____cm

Radius = _____cm

Calculate volume of **wide** aluminum cylinder:

(WRITE FORMULA:)

V =

(SHOW CALCULATIONS / UNITS / 2 DECIMAL PLACES:)

V =

(WRITE ANSWER / UNITS / 2 DECIMAL PLACES:)

V =

Grading Standard:

You will probably be graded as follows:

In each box for each set of calculations:

-1/2 point for wrong answer
-1/2 point for any omissions

Minus variable number of points for messy work.

Data Table 7 (Aluminum Density)

Calculate density of the **wide** aluminum cylinder:

(WRITE FORMULA:)

D =

(SHOW CALCULATIONS / UNITS / 2 DECIMAL PLACES:)

D =

(WRITE ANSWER / UNITS / 2 DECIMAL PLACES:)

D =

Exercise 6   Report Sheet   Name:

Igneous Rock Properties   Lab:

Take an igneous rock tray and complete the chart below using your classnotes. **Do not spend much time trying to identify the minerals in the rock, rely on shade mainly.** Obsidian isn't light-colored is it, but what else can you call it? Complete the "Minerals" box regardless of texture. Usually you will list two minerals.

Number:	Texture:	Shade:	Minerals:	Name:
1	coarse		plagioclase &	
2	coarse			
3				
4				
5				
6				
7				
8				
9	fine			
10				
11				
12				
13				
14				
15	fine	medium		
16				
17				
18				
19 (Tuff is lithified ash & cinder with pyroclastic texture)				Tuff
20	coarse		augite & olivine	

Exercise 7   Report Sheet   Name:

Rock Practice Quiz 1 (Ig)   Lab:

State the texture:

1 _____ (texture)

2 _____ (texture)

3 _____ (texture)

4 _____ (texture)

5 _____ (texture)

Identify each rock:

6 _____

7 _____

8 _____

9 _____

10 _____

11 _____

12 _____

13 _____

14 _____

15 _____

16 _____

17 _____

18 _____

19 _____

20 _____

21 _____

22 _____

23 _____

24 _____

25 _____

26 _____

27 _____

28 _____

29 _____

30 _____

Exercise 8

Sedimentary Rock Key

1. chalk

2. rock salt

3. chert

4. fossiliferous limestone

5. breccia

6. sandstone

7. fossiliferous limestone

8. conglomerate

9. lignite coal

10. jasper

11. coquina

12. limestone

13. barite rose concretion

14. agate

15. shale

16. lithographic limestone

17. flint

18. siltstone

19. rock gypsum

20. travertine

Report Sheet   Name:

Rock Practice Quiz 2 (Sed)   Lab:

Identify the sedimentary rock:

1 _____   25 _____

2 _____   26 _____

3 _____   27 _____

4 _____   28 _____

5 _____   29 _____

6 _____   30 _____

7 _____

8 _____

9 _____

10 _____

11 _____

12 _____

13 _____

14 _____

15 _____

16 _____

17 _____

18 _____

19 _____

20 _____

21 _____

22 _____

23 _____

24 _____

Exercise 9	Report Sheet	Name:

Rock Practice Quiz 3 (Ig/Sed)   Lab:

Identify the rock and state its family (I or S):

1 _____ – _____

2 _____ – _____

3 _____ – _____

4 _____ – _____

5 _____ – _____

6 _____ – _____

7 _____ – _____

8 _____ – _____

9 _____ – _____

10 _____ – _____

11 _____ – _____

12 _____ – _____

13 _____ – _____

14 _____ – _____

15 _____ – _____

16 _____ – _____

17 _____ – _____

18 _____ – _____

19 _____ – _____

20 _____ – _____

21 _____ – _____

22 _____ – _____

23 _____ – _____

24 _____ – _____

25 _____ – _____

26 _____ – _____

27 _____ – _____

28 _____ – _____

29 _____ – _____

30 _____ – _____

Exercise 10

Metamorphic Rock Key

1. slate

2. slate

3. marble

4. gneiss

5. marble

6. gneiss

7. quartzite

8. phyllite

9. hornfels

10. mica schist

11. hornblende schist

12. anthracite coal

13. quartzite

Report Sheet   Name: _____

Rock Practice Quiz 4 (Ig/Sed/Meta)   Lab: _____

Identify the rock and state its family (I, S or M):

1 _____ – _____   25 _____ – _____

2 _____ – _____   26 _____ – _____

3 _____ – _____   27 _____ – _____

4 _____ – _____   28 _____ – _____

5 _____ – _____   29 _____ – _____

6 _____ – _____   30 _____ – _____

7 _____ – _____

8 _____ – _____

9 _____ – _____

10 _____ – _____

11 _____ – _____

12 _____ – _____

13 _____ – _____

14 _____ – _____

15 _____ – _____

16 _____ – _____

17 _____ – _____

18 _____ – _____

19 _____ – _____

20 _____ – _____

21 _____ – _____

22 _____ – _____

23 _____ – _____

24 _____ – _____

Exercise 11   Report Sheet   Name:

Rock Practice Quiz 5 (Ig/Sed/Meta)   Lab:

Identify the rock and state its family (I, S or M):

1 _____ – _____   25 _____ – _____

2 _____ – _____   26 _____ – _____

3 _____ – _____   27 _____ – _____

4 _____ – _____   28 _____ – _____

5 _____ – _____   29 _____ – _____

6 _____ – _____   30 _____ – _____

7 _____ – _____

8 _____ – _____

9 _____ – _____

10 _____ – _____

11 _____ – _____

12 _____ – _____

13 _____ – _____

14 _____ – _____

15 _____ – _____

16 _____ – _____

17 _____ – _____

18 _____ – _____

19 _____ – _____

20 _____ – _____

21 _____ – _____

22 _____ – _____

23 _____ – _____

24 _____ – _____

Exercise 12   Questions   Name:

Texas Geological Highway Map   Lab:

Note: Subsurface means below the surface. Complete the Scantron:

1 _____ It is okay to write on the map.
(A) True, (B) False

2 _____ are the initials of the map's publisher.
(A) AAPG, (B) USGS, (C) BEG, (D) DGS, (E) None

There are five "Generalized Charts of Time and Rock Units" displayed around the margins of the map (henceforth called the **Big Map**). These are also called "columnar sections." (Like a cross-section in a column.) They show what rocks are found beneath the surface and at the surface. Now refer to the little box (at the bottom of the Big Map) entitled: "Areas Represented on Columnar Sections Texas." Notice that Corpus Christi is in the green area labeled: "Southwest Texas and Lower Gulf Coast." Therefore the columnar section for Corpus Christi is the one in the lower right corner of the Big Map.

Now find Beeville just north of Corpus Christi on the Big Map **(Beeville has the same columnar section)**. Notice Beeville is on the orange color with **the symbol: "Tp"**. Find "Tp" on the columnar section and notice that two formations have that symbol—they are Willis and Goliad. Beeville is built on one or both of these formations. If the Willis Formation is outcropping then the Goliad Formation is the first subsurface formation. **The higher formations on the columnar sections are younger** therefore the Willis is the youngest and the Goliad is the oldest formation outcropping at Beeville.

State the name of the columnar section for the following cities:

3 _____ – Corpus Christi (A) East Texas, (B) High Plains, (C) West Texas,
(D) Southwest Texas, (E) North Central Texas

4 _____ – San Antonio (A) East Texas, (B) High Plains, (C) West Texas,
(D) Southwest Texas, (E) North Central Texas

5 _____ – Dallas (A) East Texas, (B) High Plains, (C) West Texas,
(D) Southwest Texas, (E) North Central Texas

6 _____ – El Paso (A) East Texas, (B) High Plains, (C) West Texas,
(D) Southwest Texas, (E) North Central Texas

FOR BEEVILLE, TEXAS:

7 _____ is the name of the columnar section for Beeville. (A) East Texas,
(B) High Plains, (C) West Texas, (D) SW Texas, (E) NC Texas

8 _____ is the symbol for the rocks at Beeville.
(A) K_{ef}, (B) P_{L1}, (C) K_{au}, (D) T_p, (E) K_t

9 _____ is the youngest formation which could outcrop at Beeville.
(A) Willis, (B) Goliad, (C) Alamo, (D) Buda, (E) None

10 _____ is the oldest formation which could outcrop there.
(A) Willis, (B) Goliad, (C) Alamo, (D) Buda, (E) None

11 _____ is the first subsurface formation at Beeville if only the Willis
formation is outcropping.
(A) Willis, (B) Goliad, (C) Alamo, (D) Buda, (E) None

State the name of the columnar section for the following cities:

12 _____ – Abilene (A) East Texas, (B) High Plains, (C) West Texas,
(D) North Central Texas, (E) Southwest Texas

13 _____ – Lubbock (A) East Texas, (B) High Plains, (C) West Texas,
(D) North Central Texas, (E) Southwest Texas

14 _____ – Terlingua (A) East Texas, (B) North Central Texas,
(C) West Texas, (D) High Plains, (E) Southwest Texas

FOR EAGLE PASS, TEXAS:

15 _____ is the name of the columnar section for Eagle Pass.
(A) East Texas, (B) High Plains, (C) West Texas, (D) Southwest
Texas, (E) North Central Texas

16 _____ is a formation which might outcrop at Eagle Pass.
(A) Austin, (B) Buda, (C) Del Rio, (D) Escondido, (E) None

17 _____ is another formation which might outcrop at Eagle Pass.
(A) Choza, (B) Olmos, (C) San Miguel, (D) Arroyo, (E) None

If the Olmos Formation is outcropping at Eagle Pass then:

18 _____ is the first subsurface formation.
Again: Subsurface means below the surface.
(A) Choza, (B) Olmos, (C) San Miguel, (D) Anacacho, (E) None

19 _____ is the second subsurface formation.
(A) Choza, (B) Olmos, (C) San Miguel, (D) Anacacho, (E) None

FOR ABILENE, TEXAS:

20 _____ is the name of the columnar section for Abilene.
(A) East Texas, (B) High Plains, (C) West Texas,
(D) North Central Texas, (E) Southwest Texas

The three formations which may outcrop at Abilene are:

21 _____ is a formation which might outcrop at Abilene.
(A) Choza, (B) Olmos, (C) San Miguel, (D) Anacacho, (E) None

22 _____ is another formation which might outcrop at Abilene.
(A) Lake Kemp, (B) Olmos, (C) San Miguel, (D) Vale, (E) None

23 _____ is still another formation which might outcrop at Abilene.
(A) Lake Kemp, (B) Olmos, (C) Arroyo, (D) Antlers, (E) None

Notice the rock symbol next to the Choza Formation on the Abilene columnar section. It is mainly shale therefore the rock name would be "Shale" hence the Choza Shale is the probable formation name instead of Choza Formation since there **is** a dominant rock type. So to continue:

24 _____ is the rock name for the Vale formation.
(A) Sandstone, (B) Formation, (C) Limestone, (D) Shale, (E) None

25 _____ is the rock name for the Arroyo formation.
(A) Sandstone, (B) Formation, (C) Limestone, (D) Shale, (E) None

26 _____ is the first subsurface formation if the Arroyo formation is outcropping. (A) Lake Kemp, (B) Capps, (C) Arroyo, (D) Vale, (E) None

27 _____ is the rock name for the Lake Kemp formation.
(A) Sandstone, (B) Formation, (C) Limestone, (D) Shale, (E) None

If you were to keep drilling at Abilene you would encounter the Ricker Station Limestone (in the bright blue). Notice that the Ricker Station Limestone grades laterally into the Garner Shale. Now look above the Ricker Station Limestone for the East Mountain formation.

28 _____ is the rock name for the East Mountain formation.
(A) Sandstone, (B) Formation, (C) Limestone, (D) Shale, (E) None

29 _____ is the formation that the East mountain formation grades laterally into. (A) Lake Kemp, (B) Capps, (C) Arroyo, (D) Vale, (E) None

30 _____ is the rock name of the Capps formation.
(A) Sandstone, (B) Formation, (C) Limestone, (D) Shale, (E) Rock Gypsum

Similarly up near the top of Abilene's columnar section (in the green) the Glen Rose Limestone grades laterally into the Antlers:

31 _____ (A) Sandstone, (B) Formation, (C) Limestone, (D) Shale

FOR WHITESBORO, TEXAS (north of Dallas near the Red River):

32 _____ is the name of the columnar section. (A) East Texas, (B) High Plains, (C) West Texas, (D) Southwest Texas, (E) North Central Texas

Outcropping at Whitesboro is teither the Woodbine Sandstone or the formation which it grades laterally into which is the:

33 _____ (geographic name)
(A) Lake Kemp, (B) Capps, (C) Antlers, (D) Pepper, (E) None

34 _____ (rock name)
(A) Sandstone, (B) Formation, (C) Limestone, (D) Shale, (E) None

Read about igneous rocks at the top of the **Big Map** in the little box entitled: "Geological Highway Map":

35 _____ is the color that represents igneous rocks on this map.
(A) Green, (B) Blue, (C) Red, (D) Gray, (E) Yellow

36 _____ Now, at last, it is okay to write on the map.
(A) True, (B) False

37 _____ are what the heavy black lines on the map indicate.
(A) Highways, (B) Coal, (C) Railroads, (D) Faults, (E) None

38 _____ is the escarpment of Texas featuring heavy black lines.
(A) Oak Cliff, (B) Caprock, (C) Rimrock, (D) Balcones, (E) None

39 _____ is the color that represents igneous rocks on this map.
(A) Green, (B) Blue, (C) Red, (D) Gray, (E) Yellow

Now look at the **Big Map** itself:

40 _____ are two areas where igneous rocks outcrop in Texas.
(A) North & South, (B) Central & West, (C) East & Central, (D) North & West, (E) North & Central

Refer back to the little box entitled: "Geological Highway Map":

41 _____ is the area where the igneous rocks are **volcanic** (= extrusive). (A) North, (B) South, (C) East, (D) West, (E) Central

42 _____ is the area where the igneous rocks are **intrusive.**
(A) North, (B) South, (C) East, (D) West, (E) Central

43 _____ is the area where the igneous rocks are **lava flows.**
(A) North, (B) South, (C) East, (D) West, (E) Central
44 _____ is the area where the igneous rocks are **batholiths.**
(A) North, (B) South, (C) East, (D) West, (E) Central
45 _____ lines is the symbol which represents metamorphic rocks.
(A) Dotted, (B) Dashed, (C) Wavy

Look at the very bottom of the columnar sections:

46 _____ are the two colors used to represent metamorphic rocks (other than green perhaps):
(A) Red & Brown, (B) Red & Purple, (C) Brown & Purple
47 _____ Metamorphic rocks outcrop in central Texas.
(A) True, (B) False
48 _____ Metamorphic rocks outcrop in west Texas.
(A) True, (B) False

49 _____ is the **most** abundant rock **family** in Texas.
(A) Igneous, (B) Sedimentary, (C) Metamorphic
50 _____ is the **least** abundant rock **family** in Texas.
(A) Igneous, (B) Sedimentary, (C) Metamorphic

Refer to the little box entitled: "Index of AAPG Geological Highway Maps":
51 _____ Region is the name of the map for Florida. (A) Mid-Continents, (B) Great Lakes, (C) Southeastern, (D) Mid-Atlantic, (E) None
52 _____ Region is the name of the map for Missouri. (A) Mid-Continents, (B) Great Lakes, (C) Southeastern, (D) Mid-Atlantic, (E) None
53 _____ Region is the name of the map for Indiana. (A) Mid-Continents, (B) Great Lakes, (C) Southeastern, (D) Mid-Atlantic, (E) None

FOR PALO PINTO, TEXAS (in the bright blue west of Ft. Worth):

54 _____ is the name of the columnar section. (A) East Texas, (B) High Plains, (C) West Texas, (D) North Central Texas, (E) Southwest Texas
55 _____ is the symbol for the age of the rocks there.
(A) K_{ef}, (B) P_{L1}, (C) IP_{mi}, (D) T_p, (E) None
56 _____ is the **geographic name** for the youngest formation which might outcrop there.
(A) Eastland, (B) Ranger, (C) Salesville, (D) Home Creek
57 _____ is the **rock name** for the youngest formation which might outcrop there. (A) Sandstone, (B) Formation, (C) Limestone, (D) Shale
58 _____ is the **geographic name** for the oldest formation which might outcrop there. (A) Eastland, (B) Ranger, (C) Salesville, (D) Home Creek
59 _____ is the **rock name** for the oldest formation which might outcrop there.
(A) Sandstone, (B) Formation, (C) Limestone, (D) Shale
60 _____ is the name of the formation outcropping there which contains chert.
(A) Eastland, (B) Ranger, (C) Salesville, (D) Home Creek, (E) None

KEEP THESE QUESTIONS.
TURN-IN THE SCANTRON SHEET.

Exercise 13

Reading Topographic Maps

Topographic maps are maps which indicate the elevation of the landscape. They do so by means of contour lines. Every point on a certain contour line has the same elevation. Thus you can say that contour lines connect points of equal elevation. The elevation of some contour lines can be read directly from the map because their elevations are marked on the lines. These are the index contour lines and are **darker** and wider than the others.

The thinner contour lines are not marked and you must therefore be able to calculate their elevations. This is done by means of contour intervals. The contour interval (C.I.) is the difference in elevation between two adjacent contour lines. Thus if the contour interval is 20 feet, for example, then one contour line is 20 feet higher or lower than the one next to it. Usually the contour interval is printed on the map but you can always calculate it as follows:

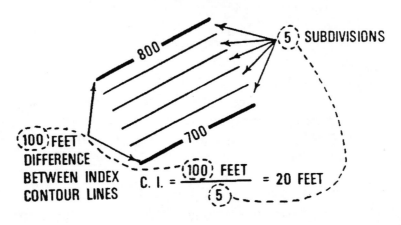

Figure 13.1

Thus in the above example the contour interval (C.I.) is 20 feet. You can always check your calculation by starting at the lower index contour line and count up to the next index contour using your calculated contour interval. Your count should agree with the printed value of the next index contour line. Here is another example of how to calculate the contour interval where the result is the C.I. = 50 feet:

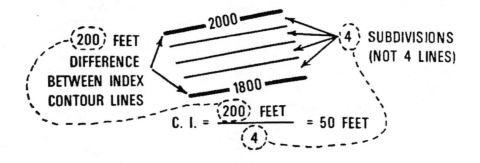

Figure 13.2

SOURCE: From *Earth, Sea, & Sky Lab Manuel* by William McLoda. Copyright © 1977 by Dallas County Community College District. Reprinted by permission.

The C.I. may vary from map to map so always check to be sure that you are using the correct value. Now let us calculate some elevations using the following example which represents the side of a hill:

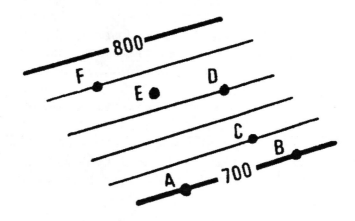

Figure 13.3

First, you must determine the C.I. It is 20 feet like before. Notice that point A and point B both lie on the index contour line which is labeled 700 feet. Therefore both points have an elevation of 700 feet. Point C lies one contour interval uphill for the index contour so its elevation is 20 feet greater or 720 feet. Point D is three contour intervals up from 700 feet or two contour intervals down from 800 (it does not matter which way you figure it), so point D's elevation is 760 feet.

Point E lies between the lines which represent 760 and 780 feet, so point E's elevation is 770 feet. (Always choose the halfway number, do not make up odd numbers like 769 or 774, pick 770 or you may be marked incorrect). Point F is one contour interval below 800 so its elevation is 780 feet.

Now let us superimpose two additional features on the above example: a small hill and a depression as if someone dug a hole and piled up the soil next to it:

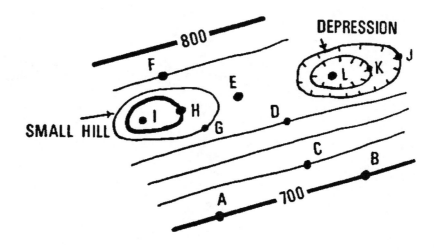

Figure 13.4

Points A to F still have the same values, of course. Point G lies one contour interval uphill from point D (which is 760 feet) so point G's elevation is 780 feet. Point H is one contour interval uphill from Point G so it is 20 feet higher or 800 feet in elevation. Point I is the top of the small hill and lies between 800 feet and the next higher contour value (which would be 820), so let us say its elevation is 810 feet (halfway again).

Now let us look at the depression. Notice that the contour lines have marks pointing inward. That is how you know it is a depression instead of another small hill. These marks are called hachure marks. Here is the **Rule of Depressions: When going uphill the first "ring" of a depression is always the same elevation as the preceding contour line.** So point J has the same value as point D. They are both 760 feet in elevation.

Now we are going downhill into the depression so point K is one contour interval less than point J and hence 740 feet. Point L lies between 740 and the next lower value (720), so let us choose the middle value again: 730 feet. Notice that the small hill and the depression have no effect on the elevation of the original points including point E and point F.

One final note about calculating elevations—suppose you can only find one index contour in a certain region of the map:

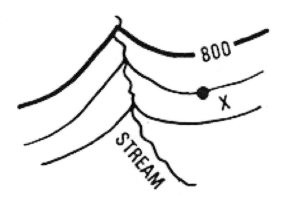

Figure 13.5

Here is the question: What is the elevation of point X? Is point X uphill or downhill from the 800 foot contour line? The Rule of V's will tell you. Here is the **Rule of V's: When contour lines meet a stream the lines "V" (point) uphill.** Therefore in the above example the 800 foot contour line lies uphill from point X so the elevation of point X is one contour interval less that 800 feet. Point X is 780 feet.

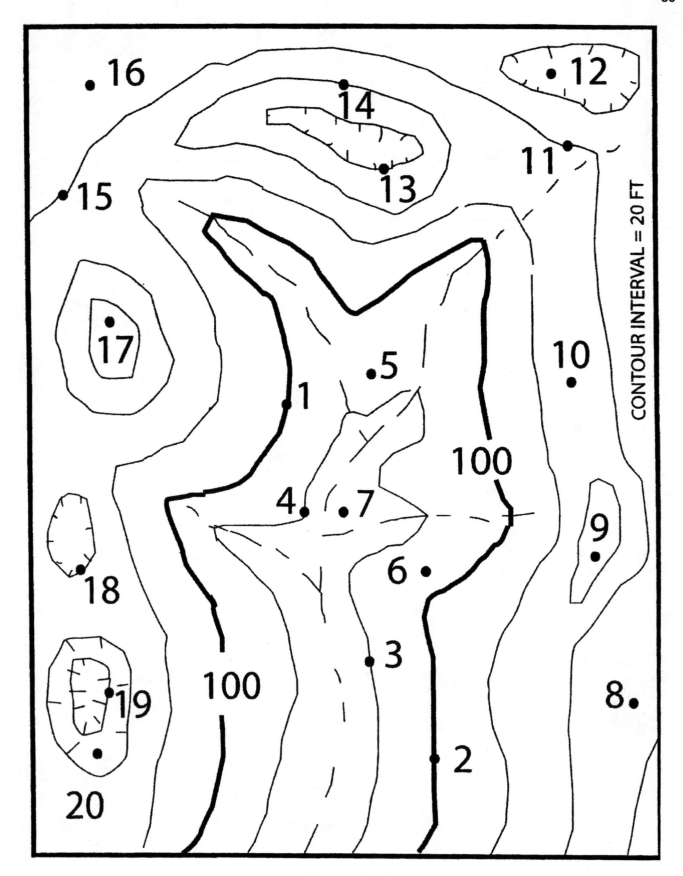

CONTOUR INTERVAL = 20 FT

Figure 13.6

39

40

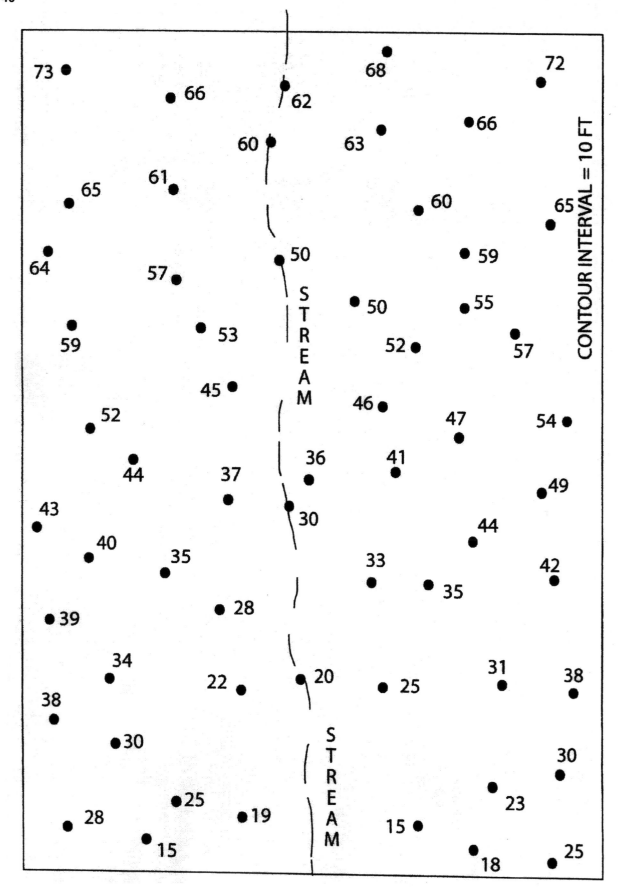

Figure 13.7

Questions Name:

Topo Maps 1 (Elevations) Lab:

1_______________ feet is the contour interval.
(A) 5, (B) 10, (C) 20, (D) 50, (E) None

State the elevation:

2_______________ feet—Point A
(A) 770, (B) 690, (C) 740, (D) 760, (E) None

3_______________ feet—Point B
(A) 770, (B) 690, (C) 740, (D) 760, (E) None

4_______________ feet—Point C
(A) 770, (B) 690, (C) 740, (D) 760, (E) None

5_______________ feet—Point D
(A) 770, (B) 690, (C) 740, (D) 760, (E) None

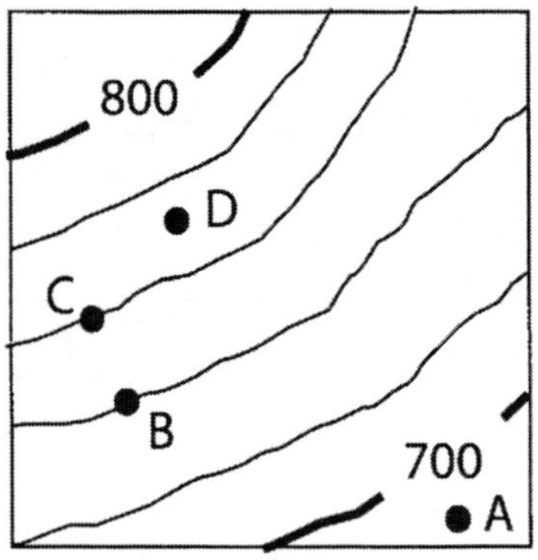

Figure 13.8

6_______________ feet is the contour interval.
(A) 5, (B) 10, (C) 20, (D) 50, (E) None

State the elevation:

7_______________ feet—Point A
(A) 85, (B) 70, (C) 55, (D) 60, (E) None

8_______________ feet—Point B
(A) 85, (B) 70, (C) 55, (D) 60, (E) None

9_______________ feet—Point C
(A) 85, (B) 70, (C) 55, (D) 60, (E) None

10_______________ feet—Point D
(A) 85, (B) 70, (C) 55, (D) 60, (E) None

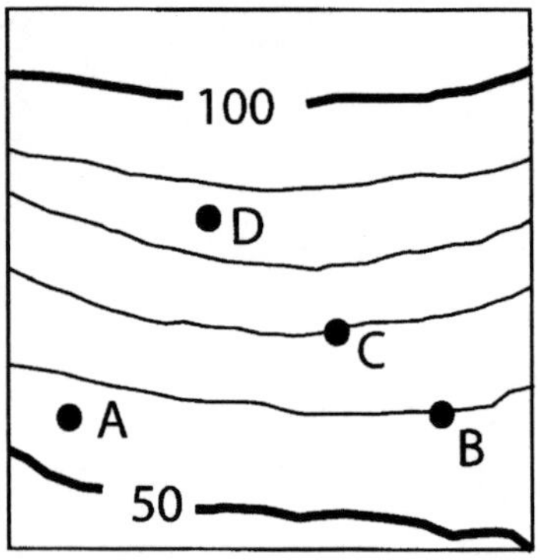

Figure 13.9

11_______________ feet is the contour interval.
(A) 5, (B) 10, (C) 20, (D) 50, (E) None

State the elevation:

12_______________ feet—Point A
(A) 325, (B) 400, (C) 300, (D) 425, (E) None

13_______________ feet—Point B
(A) 325, (B) 400, (C) 300, (D) 425, (E) None

14_______________ feet—Point C
(A) 325, (B) 400, (C) 300, (D) 425, (E) None

15_______________ feet—Point D
(A) 325, (B) 400, (C) 300, (D) 425, (E) None

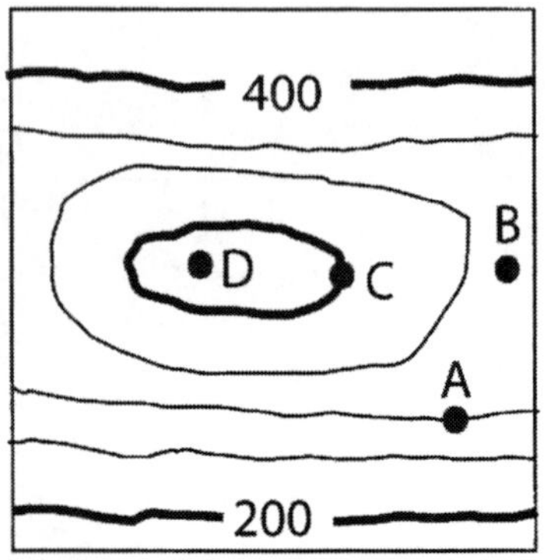

Figure 13.10

If the CI = 20 feet state the elevation:

16 _____ feet—Point A
(A) 10, (B) 0, (C) 50, (D) 20, (E) None
17 _____ feet—Point B
(A) 10, (B) 0, (C) 50, (D) 20, (E) None
18 _____ feet—Point C
(A) 10, (B) 0, (C) 50, (D) 20, (E) None
19 _____ feet—Point D
(A) 10, (B) 0, (C) 50, (D) 20, (E) None
20 _____ feet—Point E
(A) 10, (B) 0, (C) 50, (D) 20, (E) None

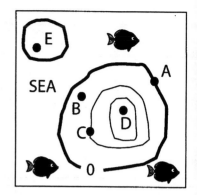

Figure 13.11

---

21 _____ is the uphill direction.
(A) East, (B) West
22 _____ is the direction the stream flows.
(A) East, (B) West

If the CI = 20 feet state the elevation:

23 _____ feet—Point A
(A) 180, (B) 190, (C) 210, (D) 220, (E) None
24 _____ feet—Point B
(A) 180, (B) 190, (C) 210, (D) 220, (E) None
25 _____ feet—Point C
(A) 180, (B) 190, (C) 210, (D) 220, (E) None

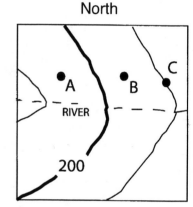

Figure 13.12

---

26 _____ is the uphill direction.
(A) Northeast, (B) Southwest
27 _____ is the direction the stream flows.
(A) Northeast, (B) Southwest

If the CI = 10 feet state the elevation:

28 _____ feet—Point A
(A) 55, (B) 45, (C) 35, (D) 65, (E) None
29 _____ feet—Point B
(A) 55, (B) 45, (C) 35, (D) 65, (E) None
30 _____ feet—Point C
(A) 55, (B) 45, (C) 35, (D) 65, (E) None

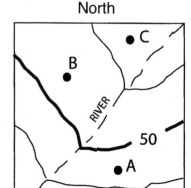

Figure 13.13

31_____ Feet is the contour interval.
(A) 5, (B) 10, (C) 20, (D) 50, (E) None

State the elevation:

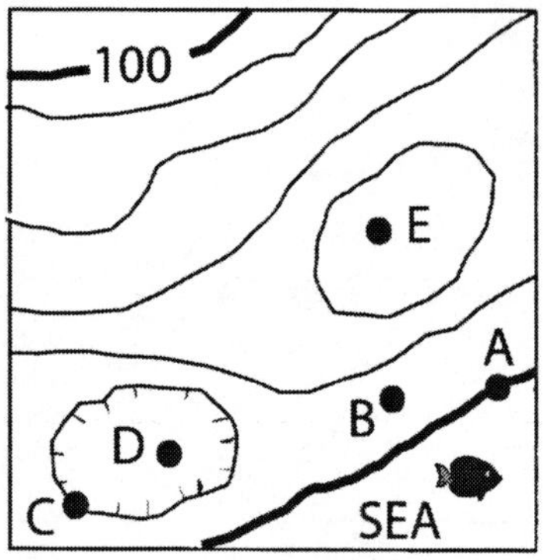

32_____ feet—Point A
(A) 10, (B) −10, (C) 0, (D) 50, (E) None
33_____ feet—Point B
(A) 10, (B) −10, (C) 0, (D) 50, (E) None
34_____ feet—Point C
(A) 10, (B) −10, (C) 0, (D) 50, (E) None
35_____ feet—Point D
(A) 10, (B) −10, (C) 0, (D) 50, (E) None
36_____ feet—Point E
(A) 10, (B) −10, (C) 0, (D) 50, (E) None

Figure 13.14

If the CI = 20 feet state the elevation:

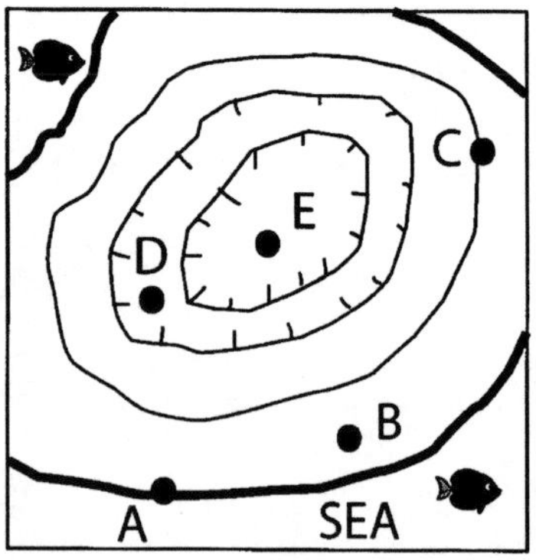

37_____ feet—Point A
(A) 5, (B) −5, (C) 0, (D) 10, (E) None
38_____ feet—Point B
(A) 5, (B) −5, (C) 0, (D) 10, (E) None
39_____ feet—Point C
(A) 5, (B) −5, (C) 0, (D) 10, (E) None
40_____ feet—Point D
(A) 5, (B) −5, (C) 0, (D) 10, (E) None
41_____ feet—Point E
(A) 5, (B) −5, (C) 0, (D) 10, (E) None

Figure 13.15

42_____ is the uphill direction.
(A) East, (B) West
43_____ is the direction the river flows.
(A) East, (B) West

If the CI = 10 feet state the elevation:

North

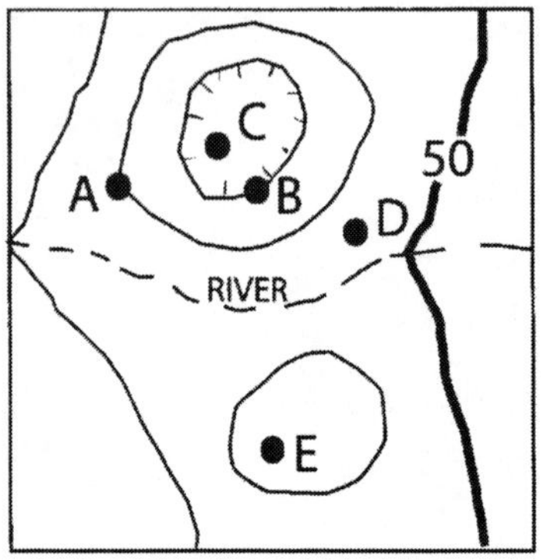

44_____ feet—Point A
(A) 65, (B) 55, (C) 50, (D) 60, (E) None
45_____ feet—Point B
(A) 65, (B) 55, (C) 50, (D) 60, (E) None
46_____ feet—Point C
(A) 65, (B) 55, (C) 50, (D) 60, (E) None
47_____ feet—Point D
(A) 65, (B) 55, (C) 50, (D) 60, (E) None
48_____ feet—Point E
(A) 65, (B) 55, (C) 50, (D) 60, (E) None

Figure 13.16

49 _____ is the contour interval.
(A) 10, (B) 20, (C) 25, (D) 50, (E) None

50 _____ is the contour line which is drawn incorrectly.
(A) 40, (B) 50, (C) 60, (D) 70, (E) None

51 _____ is the direction the stream flows.
(A) North, (B) South, (C) East, (D) West

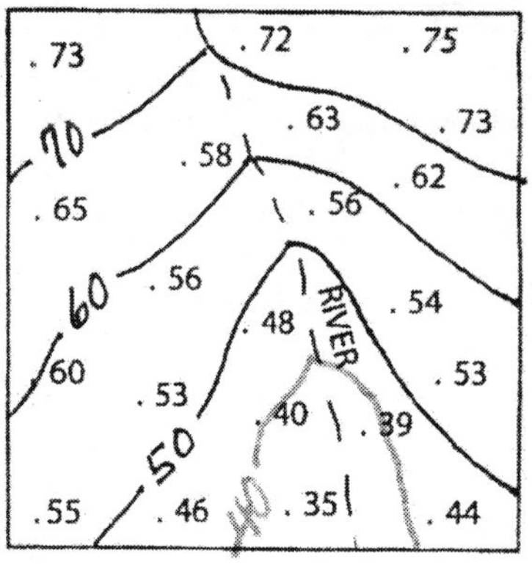

Figure 13.17

52 _____ is the contour interval.
(A) 10, (B) 20, (C) 25, (D) 50, (E) None

53 _____ is the contour line which is drawn incorrectly.
(A) 40, (B) 30, (C) 20, (D) 10, (E) None

54 _____ is the direction the stream flows.
(A) North, (B) South, (C) East, (D) West

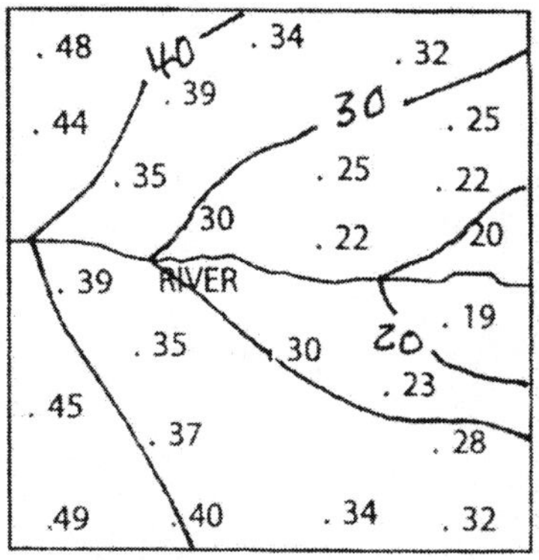

Figure 13.18

55 _____ is the contour interval.
(A) 10, (B) 20, (C) 25, (D) 50, (E) None

56 _____ is a contour line which is drawn incorrectly.
(A) 40, (B) 50, (C) 60, (D) 70, (E) None

57 _____ is the direction the stream on the upper left flows.
(A) Southwest, (B) Southeast, (C) Northeast, (D) Northwest

58 _____ is a contour line which is drawn incorrectly.
(A) 80, (B) 30, (C) 20, (D) 90, (E) None

59 _____ is the direction the stream on the upper right flows.
(A) Southwest, (B) Southeast, (C) Northeast, (D) Northwest

60 _____ is the direction the stream on the lower center flows.
(A) Southwest, (B) Southeast, (C) Northeast, (D) Northwest

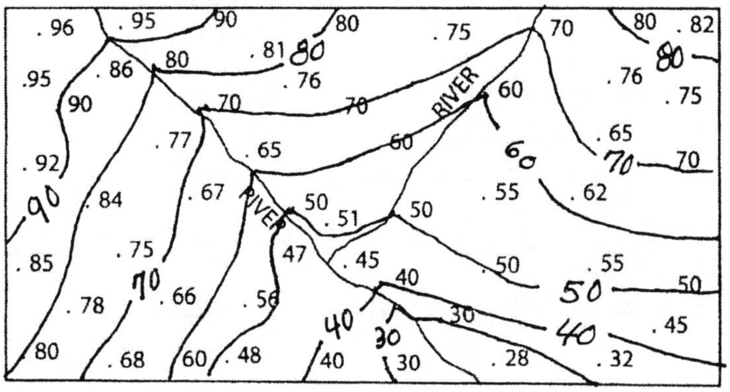

Figure 13.19

Exercise 14   Instructions

Topo Maps 2 (Quadrangles)

Take a Garland Quadrangle and see if you understand how to arrive at the following answers. When you understand, then complete the Report Sheet.

Note: 1 degree = 60 minutes, 1 minute = 60 seconds

---

1 FALSE _____ (True or False) It is okay to write on the map.
                          (Ask your instructor if you are little fuzzy about this.)
2 GARLAND QUAD _____ is the name of this map.
                          (upper right corner)
3 WHITE ROCK LAKE ____ is the map which continues south of this map.
                          (bottom center)
4 DALLAS _____   is the map which continues southwest of this map.
                              (lower left corner)
5 ROWLETT _____ is the map which continues east of this map.
                          (right center)
6 7.5 MIN _____ is the size of this map.
                          (upper right corner)
7 10 _____ feet is the contour interval.
                          (bottom center)
8 1:24,000 _____ is the scale of this map.
                          (bottom center)
9 24,000 _____ feet of actual land are represented by one foot on this map.
                          (one of anything on the map = 24,000 of actual land)
10 NORTHEAST _____ is the location of this area in Texas.
                          (black square in Texas outline at bottom)
11 32°52′30″ N _____ is the latitude of the bottom of the map.
                          (either end of the bottom edge)
12 33°N _____ is the latitude of the top of the map.
                          (either end of the top edge)
13 7.5′ _____ is the difference in latitude between the top and bottom.
                          (subtract: 33° − 32° 52′30″ − 7.5′)
14 96°37′30″ W _____ is the longitude of the right side of the map.
                          (either end of the right edge)
15 96°45′ W _____ is the longitude of the left side of the map.
                          (either end of the left edge)
16 7.5′ _____ is the difference in longitude between the two sides.
                          (subtract: 96°45′ − 96°37′ 30″ = 7.5′)

Questions   Name:

Topo Maps 2 (Quadrangles)   Lab:

PLEASE NOTE: For any points not on contour lines please write the halfway numbers.
Ex: For a point between 560 and 570 write 565.
Please read the Instruction Sheet first. Complete the Scantron form.

Garland Quadrangle:

1 _____ It is okay to write on the Garland Quadrangle.
(A) True, (B) False, (C) If one is discreet
2 _____ Quadrangle is the map which continues southeast of this map.
(A) Plano, (B) Dallas, (C) Mesquite, (D) Addison, (E) Rowlett
3 _____ Quadrangle is the map which continues east of this map.
(A) Plano, (B) Dallas, (C) Mesquite, (D) Addison, (E) Rowlett
4 _____ Quadrangle is the map which continues southwest of this map.
(A) Plano, (B) Dallas, (C) Mesquite, (D) Addison, (E) Rowlett
5 _____ feet is the contour interval.
(A) 5, (B) 10, (C) 20, (D) 50, (E) None
6 _____ is the scale of this map.
(A) 24,000; (B) 62,500; (C) 1:24,000; (D) 1:62,500
7 _____ minutes is the size of this map.
(A) 7.5, (B) 15
8 _____ is the direction which Dixon Branch (south center) flows.
(A) N, (B) S, (C) E, (D) W
9 _____ is the latitude of the bottom of this map.
(A) 96°45′W, (B) 96°45′E, (C) 96°37′30″W, (D) 32°52′30″N, (E) 32°52′30″S
10 _____ is the longitude of the west side of this map.
(A) 96°45′W, (B) 96°45′E, (C) 32°52′30″W, (D) 32°52′30″N, (E) 32°52′30″S
11 _____ is the longitude of the east side of this map.
(A) 96°45′W, (B) 96°45′E, (C) 96°37′30″W, (D) 32°52′30″N, (E) 96°37′30″E

Determine the elevation of the following locations:

12 _____ feet = Shiloh @ West Miller
(A) 565, (B) 570, (C) 575, (D) 585, (E) None
13 _____ feet = Shiloh @ West Way
(A) 565, (B) 570, (C) 575, (D) 585, (E) None
14 _____ feet = Shiloh @ Cardinal Lane
(A) 575, (B) 580, (C) 585, (D) 595, (E) None
15 _____ feet = West Way @ Hilltop
(A) 575, (B) 580, (C) 585, (D) 590, (E) None
16 _____ feet = West Way @ Patricia
(A) 565, (B) 570, (C) 575, (D) 585, (E) None
17 _____ feet = North Star @ Walnut
(A) 565, (B) 570, (C) 575, (D) 580, (E) None
18 _____ feet = North Star @ Dent St
(A) 570, (B) 575, (C) 540, (D) 550, (E) None
19 _____ feet = North Star @ Bandera
(A) 570, (B) 575, (C) 540, (D) 585, (E) None

Bray Quadrangle:

20 _____ minutes is the size of this map.
(A) 7.5, (B) 15

21 _____ inches of land equal one inch on the map.
(A) 24,000; (B) 62,500; (C) 1:24,000; (D) 1:62,500

22 _____ California is the location of this map.
(A) Northern, (B) Southern

23 _____ is the direction of flow of Antelope Creek (center).
(A) North, (B) South, (C) East, (D) West

24 _____ is the latitude of the bottom of this map.
(A) 122°E, (B) 122°W, (C) 41°30′N, (D) 41°30′S, (E) 41°30′W

25 _____ is the longitude of the west side of this map.
(A) 122°E, (B) 122°W, (C) 41°30′N, (D) 41°30′S, (E) 41°30′W

26 _____ is the longitude of the east side of this map.
(A) 122°E, (B) 122°W, (C) 41°30′N, (D) 41°30′S, (E) 121°45′W

Addison Quadrangle:

27 _____ Quadrangle is the map which continues west of this map.
(A) Plano, (B) Dallas, (C) Garland, (D) Addison, (E) Carrollton

28 _____ Quadrangle is the map which continues south of this map.
(A) Plano, (B) Dallas, (C) Garland, (D) Addison, (E) Carrollton

29 _____ feet is the C.I.
(A) 5, (B) 10, (C) 20, (D) 50, (E) 80

Determine the elevation of the following locations:

30 _____ feet = Royal @ Webbs Chapel (southwest edge)
(A) 485, (B) 495, (C) 505, (D) 515, (E) None

31 _____ feet = Royal @ Crowell
(A) 485, (B) 495, (C) 505, (D) 515, (E) None

32 _____ feet = Royal @ Cox
(A) 485, (B) 495, (C) 515, (D) 525, (E) None

33 _____ feet = Royal @ Welch
(A) 565, (B) 575, (C) 585, (D) 605, (E) None

34 _____ feet = Royal @ Inwood Road
(A) 565, (B) 575, (C) 585, (D) 605, (E) None

35 _____ feet = Royal @ Preston
(A) 565, (B) 575, (C) 585, (D) 605, (E) None

36 _____ feet = Royal @ Hillcrest Road
(A) 565, (B) 575, (C) 585, (D) 595, (E) None

37 _____ Quadrangle is the map which continues east of this map.
(A) Plano, (B) Dallas, (C) Garland, (D) Addison, (E) Rowlett

38 _____ is the latitude of the bottom of this map. (A) 96°52′30″E,
(B) 96°52′30″W, (C) 32°52′30″S, (D) 32°52′30″N, (E) 96°45′W

39 _____ is the longitude of the west side of this map. (A) 96°52′30″E,
(B) 96°52′30″W, (C) 32°52′30″S, (D) 32°52′30″N, (E) 96°45′W

40 _____ is the longitude of the east side of this map.
(A) 96°52′30″E, (B) 96°52′30″W, (C) 32°52′30″S, (D) 96°45′N, (E) 96°45′W

Bright Angel Quadrangle:

41 _____ is the map which continues east of this map.
(A) Vishnu Temple, (B) Nankoweap, (C) De Motte Park, (D) Powell Plateau,
(E) Havasupai Point

42 _____ minutes is the size of this map.
(A) 7.5, (B) 15

43 _____ inches of land equal one inch on the map.
(A) 24,000; (B) 62,500; (C) 1:24,000; (D) 1:62,500

44 _____ feet is the C.I.
(A) 5, (B) 10, (C) 20, (D) 50, (E) 80

45 _____ Arizona is the location.
(A) Northern, (B) Southern

46 _____ is the latitude of the bottom of this map.
(A) 112°E, (B) 36°S, (C) 112°W, (D) 36°N

47 _____ is the longitude of the west side of this map.
(A) 112°15'E, (B) 36°S, (C) 112°15'W, (D) 36°N, (E) 112°W

48 _____ is the longitude of the east side of this map.
(A) 112°E, (B) 36°S, (C) 112°W, (D) 36°N, (E) 112°N

49 _____ is the map which continues west of this map.
(A) Vishnu Temple, (B) Nankoweap, (C) De Motte Park, (D) Powell Plateau,
(E) Havasupai Point

50 _____ is the map which continues southeast of this map.
(A) Vishnu Temple, (B) Nankoweap, (C) De Motte Park, (D) Powell Plateau,
(E) Grandview Point

Las Vegas SW Quadrangle Questions:

Determine the elevation of the following locations:

51 _____ feet = Pebble Road @ Haven St (southeast)
(A) 2215, (B) 2225, (C) 2235, (D) 2245, (E) None

52 _____ feet = Las Vegas Bvld @ Richmar Ave
(A) 2215, (B) 2225, (C) 2235, (D) 2245, (E) None

53 _____ feet = Wigwam Ave Gilespie
(A) 2215, (B) 2225, (C) 2235, (D) 2245, (E) None

54 _____ feet = Windmill Lane @ Placid St
(A) 2205, (B) 2215, (C) 2225, (D) 2195, (E) None

55 _____ feet = Wigwam @ Placid St
(A) 2215, (B) 2225, (C) 2235, (D) 2245, (E) None

56 _____ feet = Eldorado Lane @ Placid St
(A) 2155, (B) 2165, (C) 2175, (D) 2185, (E) None

57 _____ feet = Las Vegas Blvd @ Pebble Road
(A) 2215, (B) 2225, (C) 2235, (D) 2245, (E) None

58 _____ feet = Bermuda Road @ Robindale Road
(A) 2135, (B) 2145, (C) 2155, (D) 2165, (E) None

CONTINUE ON BACK

59 _____ Quadrangle is the name of this map.

(A) Las Vegas, (B) Las Vegas SE, (C) Las Vegas SW, (D) Sloan, (E) None

60 _____ is the scale.

(A) 24,000; (B) 62,500; (C) 1:24,000; (D) 1:62,500

61 _____ is closer:

(A) Prime Meridian, (B) International Dateline

62 _____ inches of land equal one inch on this map.

(A) 24,000; (B) 62,500; (C) 1:24,000; (D) 1:62,500

63 _____ minutes is the size of this quadrangle.

(A) 7.5, (B) 15

64 _____ Nevada is the location of this map.

(A) Northern, (B) Southern

65 _____ is the map that continues east of this map.

(A) Blue Diamond SE, (B) Goodsprings, (C) Las Vegas SE, (D) Sloan

66 _____ is is the map that continues west of this map.

(A) Blue Diamond SE, (B) Goodsprings, (C) Las Vegas SE, (D) Sloan

67 _____ is the map that continues southwest of this map.

(A) Blue Diamond SE, (B) Goodsprings, (C) Las Vegas SE, (D) Sloan

68 _____ is the latitude of the bottom of this map.

(A) 36°W, (B) 115°7.5′N, (C) 115°7.5′W, (D) 36°N, (E) 115°7.5′E

69 _____ is the longitude of the east side of this map.

(A) 36°W, (B) 115°07′30″N, (C) 115°07′30″W, (D) 36°N, (E) 115°07′30″E

70 _____ is the longitude of the west side of this map.

(A) 115°15′E, (B) 115°07′30″N, (C) 115°07′30″W, (D) 115°15′W,
(E) 115°07′30″E

WRITE YOUR NAME &
WRITE TOPO MAPS 2 ON YOUR SCANTRON.

Exercise 15   Questions   Name:

Magnetism   Lab:

Read each station's information sheet. Complete the Scantron:

Station 1 (Metals & Bar Magnets)

1_____ Aluminum is attracted by a magnet.
(A) True, (B) False
2_____ Zinc is attracted by a magnet.
(A) True, (B) False
3_____ Nickel is attracted by a magnet.
(A) True, (B) False
4_____ Copper is attracted by a magnet.
(A) True, (B) False
5_____ Cobalt is attracted by a magnet.
(A) True, (B) False
6_____ Iron is attracted by a magnet.
(A) True, (B) False
7_____ Your lab partner is attracted to you.
(A) **True,** (B) False

> NORTH IS DEFINED AS THE "N" END OF OUR BAR MAGNETS

Station 2 (Bar Magnets)

8_____ is what happens when the north poles of two magnets approach each other.
(A) Attraction, (B) Repulsion
9_____ is what happens when the south poles of two magnets approach each other.
(A) Attraction, (B) Repulsion
10_____ is what happens when the north pole of one magnet approaches the south pole of another. (A) Attraction, (B) Repulsion

Station 3 (Bar Magnet versus Horseshoe Magnet)

11_____ If a bar magnet is bent into a horseshoe it becomes stronger.
(A) True, (B) False
12_____ The closer the poles, the stronger the horseshoe magnet.
(A) True, (B) False

Station 4 (Electrostatic Series Chart)

If you rough up the cat with a sheet of Saran Wrap then:

13_____ is the resulting charge of the Cat.
(A) Positive, (B) Negative
14_____ is the resulting charge of the Saran Wrap.
(A) Positive, (B) Negative
15_____ lost electrons. (A) Cat, (B) Saran Wrap
The Cat's fur will then be all "sticky-outy" because like charges:
16_____ (A) Attract, (B) Repel

Station 5 (Alnico Magnet Chart)

17_____ is the main metal in Alnico magnets. (Hint: Read the dark print.)
(A) Iron, (B) Cobalt, (C) Nickel, (D) Aluminum, (E) Copper
18_____ is the second main metal in Alnico magnets.
(A) Iron, (B) Cobalt, (C) Nickel, (D) Aluminum, (E) Copper
19_____ is the third main metal in Alnico magnets.
(A) Iron, (B) Cobalt, (C) Nickel, (D) Aluminum, (E) Copper

Station 6 (Induction Device & Bar Magnet)

20_____ A moving (changing) magnetic field causes (induces) current to flow in a wire.
(A) True, (B) False
21_____ is the direction the ammeter needle deflects when the North pole is inserted.
(A) Right, (B) Left
22_____ (Right or Left) is the direction the ammeter needle deflects when the South pole
is inserted. (A) Right, (B) Left
23_____ The more coils of wire, the more the ammeter needle is deflected.
(A) True, (B) False
24_____ is the direction the ammeter needle deflects when the North pole is removed.
(A) Right, (B) Left
25_____ is the direction the ammeter needle deflects when the South pole is removed.
(A) Right, (B) Left

Station 7 (Silva Compass & Global Map & Bar Magnet)

26_____ pole is the red end of the Silva compass assuming the pole marked "N" on the
bar magnet is a north pole. (A) North, (B) South
27_____ The earth's magnetic poles and geographic poles coincide.
(A) True, (B) False
28_____ magnetic pole of earth is closest to the North geographic pole.
(A) North, (B) South
29_____ magnetic pole of earth is closest to the South geographic pole.
(A) North, (B) South

Station 8 (Penguin Photo)

30_____ is the closest geographic pole.
(A) North, (B) South
31_____ is the closest magnetic pole.
(A) North, (B) South

Station 9 (Polar Bear Photo)

32_____ is the closest geographic pole.
(A) North, (B) South
33_____ is the closest magnetic pole.
(A) North, (B) South

Station 10 (Declination)

34_____ is the name for the angle **between** lines drawn from a place to the magnetic pole and to the geographic pole.
(A) Declination, (B) Inclination

35_____ Declination varies with location.
(A) True, (B) False

36_____ is the name of the line of zero declination.
(A) Prime Meridian, (B) Equator, (C) Arctic Circle, (D) Agonic, (E) None

37_____ If you stand on the Agonic line your compass will point to the Earth's North Geographic pole. (A) True, (B) False

Station 11 (Earth's Magnetic Field)

38_____ The Earth's magnetic field is thought to be formed in the:
(A) Crust, (B) Mantle, (C) Outer Core, (D) Inner Core, (E) None

39_____ Moving charged particles create a magnetic field.
(A) True, (B) False

40_____ Stationary charged particles create a magnetic field.
(A) True, (B) False

41_____ is the part of the Earth which is molten.
(A) Crust, (B) Mantle, (C) Outer Core, (D) Inner Core, (E) None

42_____ The rotation of the Earth sets the outer core liquid in motion.
(A) True, (B) False

Station 12 (Room Map & Bar Magnet)

43_____ is the north wall of this room.
(A) A, (B) B, (C) C, (D) D

Station 13 (Inclination Diagrams)

44_____ is the name for the angle between a compass needle (tilted sideways) and the horizontal ground surface.
(A) Declination, (B) Inclination

45_____ would be the latitude of **greatest** inclination.
(A) 0, (B) 30, (C) 60, (D) 90

46_____ would be the latitude of **least** inclination.
(A) 0, (B) 30, (C) 60, (D) 90

Station 14 (Cow Magnets)

Cow magnets are sometimes given to cows to hold wire, etc. (which they accidentally swallowed) in the stomach until it dissolves because the wire could punch a hole in their stomach lining later.

47_____ are the people who use cow magnets (**Go to magnetsource.com**)
(A) cattle ranchers, (B) dairy farmers

48_____ disease is the name for metal related health problems of cows. (INTERNET)
(A) Wire, (B) Nail, (C) Hardware, (D) Metal, (E) None

Station 15 (Silva Compass & Bar Magnet)

49_____ is the north pole of this compass assuming the bar magnet is labeled correctly.
(A) Red, (B) White

Station 16 (Mineral & Bar Magnet)

50_____ is the name for this mineral.
(A) Hematite, (B) Magnetite, (C) Pyrite, (D) Siderite, (E) None

51_____ This mineral has a north pole and a south pole similar to a bar magnet.
(A) True, (B) False

Station 17 (Textbook Illustration)

52_____ This illustration out of our textbook has the magnetic poles correctly labeled.
(A) True, (B) False

Station 18 (Bar Magnets Flux Pattern)

53_____ is the result when like poles face each other.
(A) Attraction, (B) Repulsion

Station 19 (Bar Magnets Flux Pattern)

54_____ is the result when unlike poles face each other.
(A) Attraction, (B) Repulsion

Station 20 (Bar Magnets Flux Pattern)

55_____ is the type of unmarked pole.
(A) North, (B) South

Station 21 (Bar Magnets Flux Pattern)

56_____ is the type of unmarked pole.
(A) North, (B) South

Station 22 (Bar Magnets Flux Pattern)

57_____ is the type of unmarked pole.
(A) North, (B) South

Station 23 (Comparisons of Magnetic Fields)

58_____ A solenoid (coil of wire) with current flowing through it has a magnetic field similar to a bar magnet. (A) True, (B) False

59_____ A solenoid (coil of wire) without current flowing through it has a magnetic field similar to a bar magnet. (A) True, (B) False

60_____ A solenoid (coil of wire) with current flowing through it is called an electromagnet. (A) True, (B) False

COMPLETE THE SCANTRON SHEET AND TURN IT IN. KEEP THESE QUESTIONS.

Exercise 16   Instructions

Sand Grains

For best results peruse these Instructions before attempting to complete the Report Sheet:

Sediment Type (Lithogenous, Biogenous or Hydrogenous)

Lithogenous—means the sand grains have a **litho (= rock)** origin. Remember the continents are basically granite and if granite is weathered then it breaks apart into individual mineral grains like quartz, orthoclase, biotite, etc. So—if a sample appears to be little rocks **or** different minerals **or** the same mineral, especially quartz, then it is lithogenous.

Biogenous—means the sand grains have a **bio (= life)** origin. If the sand appears to be mainly complete little shells or pieces of larger shells, then the sample is biogenous. Ocean islands tend to have biogenous sand. Remember ocean islands are not granite like the continents. Ocean islands tend to be volcanic so sometimes the sand **is** lithogenous—like obsidian or olivine but usually the sea critters inundate the island's beaches with their shell debris to make the sand biogenous.

Hydrogenous—means the sand grains have a **hydro (= water)** origin. When water evaporates the chemicals dissolved in the water may be precipitated out to form little sand size mineral crystals as well as ooids (= oolites) on rare occasions. So if the sand sample has little crystals (regular shapes), not broken fragments like glass, then it is probably hydrogenous. Similarly if the sand grains all look just alike and are rounded like jelly beans **but not clear,** the sand probably consists of oolites. In hydrogenous samples the grains all tend to be homogenous (= look alike).

Composition (Homogeneous or Heterogeneous)

Homogeneous—means the sand grains are the same (homo = same) and therefore consist of the same minerals for the most part. Samples consisting mainly of the same mineral are homogeneous.

Heterogeneous—means the sand grains are different (hetero = different) and therefore consists of a variety of different minerals for the most part. Samples where a large number of sand grains **(not a few)** are different from the rest are heterogeneous.

Sorting (Sorted or Unsorted)

Sorting refers to the relative size of sand grains. Sorted sands display less variety in size—the grains are close to the same size and thus display evidence of much sorting. Unsorted sands display many different sizes.

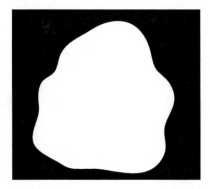

Sorted
Figure 16.1

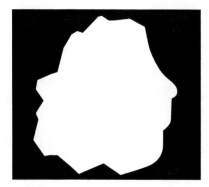

Unsorted
Figure 16.2

Rounding (Rounded or Unrounded)

Rounding refers to the relative smoothness of sand grains. Rounded grains are basically smooth as a result of having traveled far from their source. Unrounded grains show little evidence of rounding which indicates they have not traveled far.

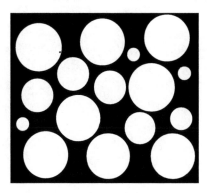

Rounded
Figure 16.3

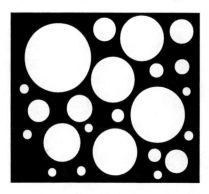

Unrounded
Figure 16.4

Questions Sheet   Name:

Sand Grains   Lab:

Please do not touch the sand, just look!

STATION 1:

1_____is the location. (A) island, (B) continent
2_____-genous is the sediment type. (A) Litho, (B) Bio, (C) Hydro
3_____is the composition. (A) Homogeneous, (B) Heterogeneous
4_____is the degree of sorting. (A) Sorted, (B) Unsorted
5_____is the degree of rounding. (A) Rounded, (B) Unrounded

STATION 2:

6_____is the location. (A) island, (B) continent
7_____-genous is the sediment type. (A) Litho, (B) Bio, (C) Hydro
8_____is the composition. (A) Homogeneous, (B) Heterogeneous
9_____is the degree of sorting. (A) Sorted, (B) Unsorted
10_____is the degree of rounding. (A) Rounded, (B) Unrounded

STATION 3:

11_____is the location. (A) island, (B) continent
12_____-genous is the sediment type. (A) Litho, (B) Bio, (C) Hydro
13_____is the composition. (A) Homogeneous, (B) Heterogeneous
14_____is the degree of sorting. (A) Sorted, (B) Unsorted
15_____is the degree of rounding. (A) Rounded, (B) Unrounded

STATION 4:

16_____is the location. (A) island, (B) continent
17_____-genous is the sediment type. (A) Litho, (B) Bio, (C) Hydro
18_____is the composition. (A) Homogeneous, (B) Heterogeneous
19_____is the degree of sorting. (A) Sorted, (B) Unsorted
20_____is the degree of rounding. (A) Rounded, (B) Unrounded

STATION 5:

21_____is the location. (A) island, (B) continent
22_____-genous is the sediment type. (A) Litho, (B) Bio, (C) Hydro
23_____is the composition. (A) Homogeneous, (B) Heterogeneous
24_____is the degree of sorting. (A) Sorted, (B) Unsorted
25_____is the degree of rounding. (A) Rounded, (B) Unrounded

STATION 6:

26_____is the location. (A) island, (B) continent
27_____-genous is the sediment type. (A) Litho, (B) Bio, (C) Hydro
28_____is the composition. (A) Homogeneous, (B) Heterogeneous
29_____is the degree of sorting. (A) Sorted, (B) Unsorted
30_____is the degree of rounding. (A) Rounded, (B) Unrounded

STATION 7:

31_____is the location. (A) island, (B) continent
32_____-genous is the sediment type. (A) Litho, (B) Bio, (C) Hydro
33_____is the composition. (A) Homogeneous, (B) Heterogeneous
34_____is the degree of sorting. (A) Sorted, (B) Unsorted
35_____is the degree of rounding. (A) Rounded, (B) Unrounded

STATION 8:

36_____is the location. (A) island, (B) continent
37_____-genous is the sediment type. (A) Litho, (B) Bio, (C) Hydro
38_____is the composition. (A) Homogeneous, (B) Heterogeneous
39_____is the degree of sorting. (A) Sorted, (B) Unsorted
40_____is the degree of rounding. (A) Rounded, (B) Unrounded

STATION 9:

41_____is the location. (A) island, (B) continent
42_____-genous is the sediment type. (A) Litho, (B) Bio, (C) Hydro
43_____is the composition. (A) Homogeneous, (B) Heterogeneous
44_____is the degree of sorting. (A) Sorted, (B) Unsorted
45_____is the degree of rounding. (A) Rounded, (B) Unrounded

STATION 10:

46_____is the location. (A) island, (B) continent
47_____-genous is the sediment type. (A) Litho, (B) Bio, (C) Hydro
48_____is the composition. (A) Homogeneous, (B) Heterogeneous
49_____is the degree of sorting. (A) Sorted, (B) Unsorted
50_____is the degree of rounding. (A) Rounded, (B) Unrounded

STATION 11:

51_____is the location. (A) island, (B) continent
52_____-genous is the sediment type. (A) Litho, (B) Bio, (C) Hydro
53_____is the composition. (A) Homogeneous, (B) Heterogeneous
54_____is the degree of sorting. (A) Sorted, (B) Unsorted
55_____is the degree of rounding. (A) Rounded, (B) Unrounded

STATION 12:

56_____is the location. (A) island, (B) continent
57_____-genous is the sediment type. (A) Litho, (B) Bio, (C) Hydro
58_____is the composition. (A) Homogeneous, (B) Heterogeneous
59_____is the degree of sorting. (A) Sorted, (B) Unsorted
60_____is the degree of rounding. (A) Rounded, (B) Unrounded

STATION 13:

61_____is the location. (A) island, (B) continent
62_____-genous is the sediment type. (A) Litho, (B) Bio, (C) Hydro
63_____is the composition. (A) Homogeneous, (B) Heterogeneous
64_____is the degree of sorting. (A) Sorted, (B) Unsorted
65_____is the degree of rounding. (A) Rounded, (B) Unrounded

STATION 14:

66_____is the location. (A) island, (B) continent
67_____-genous is the sediment type. (A) Litho, (B) Bio, (C) Hydro
68_____is the composition. (A) Homogeneous, (B) Heterogeneous
69_____is the degree of sorting. (A) Sorted, (B) Unsorted
70_____is the degree of rounding. (A) Rounded, (B) Unrounded

STATION 15:

71_____is the location. (A) island, (B) continent
72_____-genous is the sediment type. (A) Litho, (B) Bio, (C) Hydro
73_____is the composition. (A) Homogeneous, (B) Heterogeneous
74_____is the degree of sorting. (A) Sorted, (B) Unsorted
75_____is the degree of rounding. (A) Rounded, (B) Unrounded

COMPLETE THE SCANTRON AND TURN IT IN. KEEP THE QUESTIONS SHEET.

Formations

One famous formation is featured at each station. Decide which formation and fill in the blanks by referring to the information sheets with similar station numbers placed around the room. It would be helpful to **carry your class notes with you.**

These can be done in any order except: **STATION 12 MUST BE DONE LAST.**

How to read "The Lexicon":

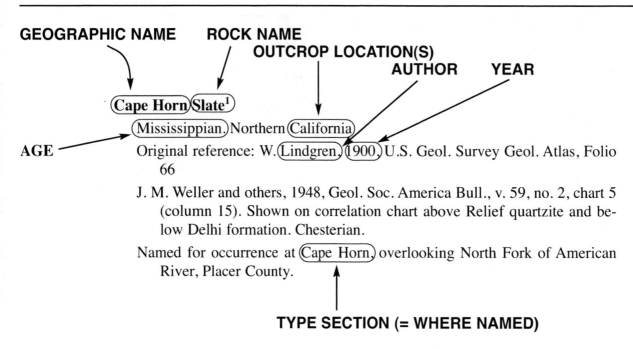

Questions   Name:

Formations   Lab:

One famous formation is featured at each station. Decide which formation and complete the SCANTRON sheet by referring to the information sheets with similar station numbers placed around the room. It would be helpful to **carry your class notes with you.**

These can be done in any order except: **STATION 12 MUST BE DONE LAST.**

STATION 1

1_____ is the **geographic** name of the formation outcropping in the bottom of the Grand Canyon.
(A) Schist, (B) St. Peter, (C) Green River, (D) Vishnu, (E) None

2_____ is the **rock** name of this formation.
(A) Formation, (B) Limestone, (C) Sandstone, (D) Schist, (E) None

3_____ is the person (last name) who is the author.
(A) Gregory, (B) Walcott, (C) Stenzel, (D) Powell, (E) None

4_____ was the year the formation was named.
(A) 1896, (B) 1932, (C) 1889, (D) 1859, (E) None

5_____ This formation outcrops in New Mexico also.
(A) True, (B) False

STATION 2

6_____ is the **geographic** name of this formation.
(A) Formation, (B) St. Peter, (C) Green River, (D) Vishnu, (E) None

7_____ is the **rock** name of this formation.
(A) Formation, (B) Limestone, (C) Sandstone, (D) Schist, (E) None

8_____ was the year this formation was named.
(A) 1896, (B) 1932, (C) 1889, (D) 1859, (E) None

9_____ is the type of vertebrate fossil for which this formation is most famous.
(A) Fish, (B) Dinosaurs, (C) Clams, (D) Turtles, (E) None

10_____ The Green River **Basin** is the only basin containing this formation.
(A) True, (B) False

STATION 3

11_____ is the **geographic** name of this formation.
(A) Formation, (B) St. Peter, (C) Green River, (D) Vishnu, (E) None

12_____ is the **rock** name of this formation.
(A) Formation, (B) Limestone, (C) Sandstone, (D) Schist, (E) None

13_____ U.S. is where this formation is located.
(A) Western, (B) Central, (C) Eastern

14_____ is the mineral that the sand grains are made of.
(A) Gypsum, (B) Feldspar, (C) Quartz, (D) Obsidian, (E) None

15_____ is the material which is manufactured from this formation.
(A) Glass, (B) Porcelain, (C) Oil, (D) Wall board, (E) None

64

STATION 4

16_____ is the **geographic** name of this formation.
(A) Bedford, (B) St. Peter, (C) Green River, (D) Vishnu, (E) None
17_____ is the **rock** name of this formation.
(A) Formation, (B) Limestone, (C) Sandstone, (D) Schist, (E) None
18_____ Limestone is the commercial name of this limestone.
(A) Cordova Shell, (B) Indiana Stone, (C) Louisville Stone
19_____ is the age (geologic period) of this formation.
(A) Jurassic, (B) Triassic, (C) Permian, (D) Mississippian, (E) None
20_____ is the type of limestone of this formation.
(A) Fossiliferous, (B) Lithographic, (C) Chalk, (D) Oolitic, (E) None

STATION 5

21_____ is the **geographic** name of this formation.
(A) Limestone, (B) Navajo, (C) Chinle, (D) Cedar Park, (E) None
22_____ is the **rock** name of this formation.
(A) Formation, (B) Limestone, (C) Sandstone, (D) Schist, (E) None
23_____ is the type of limestone of this formation.
(A) Fossiliferous, (B) Lithographic, (C) Chalk, (D) Oolitic, (E) None
24_____ is the commercial name for this formation.
(A) Cordova shell, (B) Indiana Stone, (C) Louisville Stone
25_____ is the Texas monument which is made of this building stone.
(A) San Angelo, (B) Alamo, (C) San Jacinto, (D) Washington,
(E) None

STATION 6

26_____ is the name of the island chain of which Rat Island is a member.
(A) Aleutian, (B) Hawaiian, (C) Antilles
27_____ is the **geographic** name of our favorite formation on Rat Island.
(A) Rat, (B) St. Peter, (C) Green River, (D) Vishnu, (E) None
28_____ is the **rock** name of our favorite formation on Rat Island.
(A) Formation, (B) Limestone, (C) Sandstone, (D) Schist, (E) None
29_____ was the rats' preferred mode of transportation to this location.
(A) Ship, (B) Plane, (C) Driftwood, (D) Truck, (E) None
30_____ was the year this formation was named.
(A) 1896, (B) 1932, (C) 1889, (D) 1859, (E) None

STATION 7

31_____ is the **geographic** name of the formation that forms the "High Red Wall" of the Grand Canyon.
(A) Town Mt, (B) Navajo, (C) Chinle, (D) Cedar Park, (E) None
32_____ is the **rock** name of the formation that forms the "High Red Wall" of the Grand Canyon.
(A) Formation, (B) Limestone, (C) Sandstone, (D) Schist, (E) None
33_____ is the age (geologic period) of this formation.
(A) Jurassic, (B) Triassic, (C) Permian, (D) Mississippian, (E) None

34 _____ was the year this formation was named.
(A) 1896, (B) 1932, (C) 1889, (D) 1859, (E) None

35 _____ This formation outcrops in New Mexico also.
(A) True, (B) False

STATION 8

36 _____ is the **geographic** name of this formation.
(A) Town Mt., (B) Navajo, (C) Chinle, (D) Cedar Park, (E) None

37 _____ is the **rock** name of this.
(A) Formation, (B) Limestone, (C) Sandstone, (D) Schist, (E) None

38 _____ Precambrian is the age of this formation.
(A) True, (B) False

39 _____ is the person (last name) who is the author.
(A) Gregory, (B) Walcott, (C) Stenzel, (D) Powell, (E) None

40 _____ was the year this formation was named.
(A) 1896, (B) 1932, (C) 1889, (D) 1859, (E) None

STATION 9

41 _____ is the **geographic** name of the formation.
(A) Sandstone, (B) Navajo, (C) Chinle, (D) Cedar Park, (E) None

42 _____ is the **rock** name of the formation.
(A) Formation, (B) Limestone, (C) Sandstone, (D) Schist, (E) None

43 _____ was where the sand accumulated to form this sandstone.
(A) Sea, (B) Desert

44 _____ is the person (last name) who is the author.
(A) Gregory, (B) Walcott, (C) Stenzel, (D) Powell, (E) None

45 _____ is the state where most of this formation is located.
(A) New Mexico, (B) Utah, (C) Arizona, (D) Indiana, (E) None

STATION 10

46 _____ is the **geographic** name of the formation.
(A) Town Mt., (B) Navajo, (C) Chinle, (D) Morrison, (E) None

47 _____ is the **rock** name of the formation.
(A) Formation, (B) Limestone, (C) Sandstone, (D) Schist, (E) None

48 _____ was the year this formation was named.
(A) 1896, (B) 1932, (C) 1889, (D) 1859, (E) None

49 _____ is the type of fossil for which this formation is most famous.
(A) Fish, (B) Dinosaurs, (C) Clams, (D) Turtles, (E) None

50 _____ is the age (geologic period) of this formation.
(A) Jurassic, (B) Triassic, (C) Permian, (D) Mississippian, (E) None

STATION 11

51 _____ is the **geographic** name of this formation.
(A) Town Mt., (B) Navajo, (C) Chinle, (D) Cedar Park, (E) None

52 _____ is the **rock** name of this formation.
(A) Formation, (B) Limestone, (C) Sandstone, (D) Schist, (E) None

53 _____ is the person (last name) who is the author.
(A) Gregory, (B) Walcott, (C) Stenzel, (D) Powell, (E) None

54 _____ is the state where Petrified Forest National Park is located.
(A) New Mexico, (B) Utah, (C) Arizona, (D) Indiana, (E) None

55 _____ is the type of tree in the park that is petrified.
(A) Conifer, (B) Palm, (C) Redwood, (D) Cypress, (E) None

STATION 12

REFER TO THE ILLUSTRATIONS WITH MATCHING QUESTION NUMBERS TO ANSWER WHICH FORMATION IS ASSOCIATED WITH THAT ILLUSTRATION. **SOME FORMATIONS MAY BE USED MORE THAN ONCE FOR ANSWERS!**

12-56 _____

(A) Bedford, (B) Morrison, (C) Green River, (D) Town Mt., (E) None

12-57 _____

(A) Chinle, (B) Navajo, (C) Cedar Park, (D) Rat, (E) None

12-58 _____

(A) Chinle, (B) Navajo, (C) Cedar Park, (D) Rat, (E) None

12-59 _____

(A) St. Peter, (B) Redwall, (C) Vishnu, (D) Bedford, (E) None

12-60 _____

(A) St. Peter, (B) Redwall, (C) Cedar Park, (D) Rat, (E) None

12-61 _____

(A) Chinle, (B) Navajo, (C) Cedar Park, (D) Rat, (E) None

12-62 _____

(A) St. Peter, (B) Redwall, (C) Vishnu, (D) Bedford, (E) None

12-63 _____

(A) Bedford, (B) Morrison, (C) Green River, (D) Town Mt., (E) None

12-64 _____

(A) Bedford, (B) Morrison, (C) Green River, (D) Town Mt., (E) None

12-65 _____

(A) St. Peter, (B) Redwall, (C) Vishnu, (D) Bedford, (E) None

12-66 _____

(A) Chinle, (B) Navajo, (C) Cedar Park, (D) Rat, (E) None

12-67 _____

(A) Bedford, (B) Morrison, (C) Green River, (D) Town Mt., (E) None

12-68 _____

(A) Bedford, (B) Morrison, (C) Green River, (D) Town Mt., (E) None

12-69 _____

(A) Chinle, (B) Navajo, (C) Cedar Park, (D) Rat, (E) None

12-70 _____

(A) St. Peter, (B) Redwall, (C) Vishnu, (D) Bedford, (E) None

COMPLETE THE SCANTRON AND ONLY TURN IT IN.

BE GENTLE WITH THE FOSSILS – THEY ARE EXPENSIVE!

1. brachiopod

2. pelecypod

3. sponge

4. horn coral

5. pelecypod

6. colonial coral

7. bryozoan

8. bryozoan

9. brachiopod

10. petrified wood

11. brachiopod

12. pelecypod

13. pelecypod

14. pelecypod

15. pelecypod

16. fusulinid

17. gastropod

18. cephalopod

19. brachiopod

20. scaphopod

21. cephalopod

22. graptolites

23. crinoid

24. blastoid

25. echinoid

26. echinoid

27. trilobite

28. gastropod

29. cephalopod

30. pelecypod

31. gastropod

32. brachiopod

33. worm tube(s)

34. sponge

35. _____

Fossil Practice Quiz 1   Lab:

Name each fossil:

1_____   24 _____

2_____   25 _____

3_____   26 _____

4_____   27 _____

5_____   28 _____

6_____   29 _____

7_____   30 _____

8_____

9_____

10_____

11_____

12_____

13_____

14_____

15_____

16_____

17_____

18_____

19_____

20_____

21_____

22_____

23_____

Exercise 19   Report Sheet   Name:

Fossil Practice Quiz 2   Lab:

Name each fossil:

1 _____   24 _____

2 _____   25 _____

3 _____   26 _____

4 _____   27 _____

5 _____   28 _____

6 _____   29 _____

7 _____   30 _____

8 _____

9 _____

10 _____

11 _____

12 _____

13 _____

14 _____

15 _____

16 _____

17 _____

18 _____

19 _____

20 _____

21 _____

22 _____

23 _____

Exercise 20   Report Sheet   Name:

Fossil Practice Quiz 3   Lab:

Name each fossil:

1_____   24 _____

2_____   25 _____

3_____   26 _____

4_____   27 _____

5_____   28 _____

6_____   29 _____

7_____   30 _____

8_____

9_____

10_____

11_____

12_____

13_____

14_____

15_____

16_____

17_____

18_____

19_____

20_____

21_____

22_____

23_____

Age Dating 1

1. There are ten boxes placed around the room. Leave the boxes where they are. Each box contains three fossils. Assume that all three fossils were found in the same bed (layer of rock). The question for you to answer is: What is the age of the bed? In other words: During what geologic period was the bed deposited?

2. To answer this question take each fossil, one at a time, not the whole box full, and identify it by common name. Then look up its genus name and geologic range and record all three: common name, genus and geologic range on the Report Sheet. Here is an example. If the bed contains these three fossils:

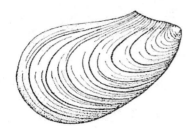

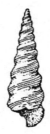

Inoceramus (J-K)

Turritella (Ħ-Q)

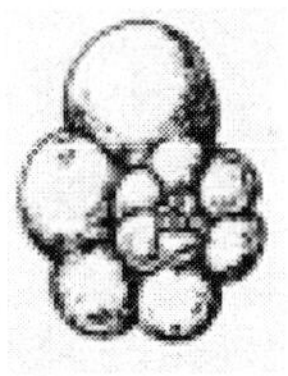

Globigerina (K-Q)

Figure 21.1

Figure 21.2

Figure 21.3

Then your Report Sheet would look like this:

Bed 0: common name / genus	Є	O	S	D	M	IP	P	Ħ	J	K	T	Q
pelecypod / Inoceramus									X	X		
gastropod / Turritella								X	X	X	X	X
foraminifera / Globigerina										X	X	X

3. Notice that the only period in which all three of the fossils lived was the Cretaceous, therefore the age of the bed is Cretaceous. Indicate this by circling the "K" column. Note: If your answer has two periods then you did something wrong; only one period is correct.

4. The number written on the fossils is the bed number. Please return each fossil to its bed. Complete the Report Sheet.

Report Sheet   Name:

Age Dating 1   Lab:

Read the Instruction Sheet and then complete Beds 1 to 10 using Bed 0 as an example:

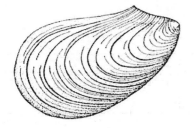

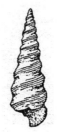

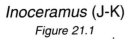

Inoceramus (J-K)
Figure 21.1

Turritella (Ƭ-Q)
Figure 21.2

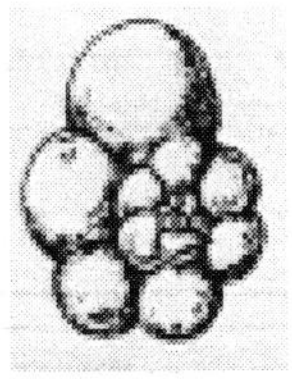

Globigerina (K-Q)
Figure 21.3

Bed 0: common name / genus	Є	O	S	D	M	IP	P	Ƭ	J	K	T	Q
pelecypod / Inoceramus									X	X		
gastropod / Turritella									X	X	X	X
foraminifera / Globigerina										X	X	X
Bed 1: common name / genus	Є	O	S	D	M	IP	P	Ƭ	J	K	T	Q

Bed 2: common name / genus   Є   O   S   D   M   IP   P   Ƭ   J   K   T   Q

Bed 3: common name / genus   Є   O   S   D   M   IP   P   Ƭ   J   K   T   Q

Bed 4: common name / genus   € O S D M IP P Ŧ J K T Q

_____

_____

_____

Bed 5: common name / genus   € O S D M IP P Ŧ J K T Q

_____

_____

_____

Bed 6: common name / genus   € O S D M IP P Ŧ J K T Q

_____

_____

_____

Bed 7: common name / genus   € O S D M IP P Ŧ J K T Q

_____

_____

_____

Bed 8: common name / genus   € O S D M IP P Ŧ J K T Q

_____

_____

_____

Bed 9: common name / genus   € O S D M IP P Ŧ J K T Q

_____

_____

_____

Bed 10: common name / genus   € O S D M IP P Ŧ J K T Q

_____

_____

_____

Exercise 22 | Questions | Name:

Age Dating 2 | | Lab:

LT means less than. GT means greater than. Fill in the blanks. The questions below are for the figures on pages 81, 82, 83 and 84:

1_____ (Granite or Sandstone) is older. (Figure 22.1)

2_____ (100, GT 100 or LT 100) million years is the age of the sandstone if the granite is 100 million years old. (Figure 22.1)

3_____ (True or False) The wavy line represents an unconformity. (Figure 22.1)

4_____ (Granite or Sandstone) is older. (Figure 22.2)

5_____ (100, GT 100 or LT 100) million years is the age of the sandstone if the granite is 100 million years old. (Figure 22.2)

6_____ is the type of unconformity represented by the wavy line. (Figure 22.2)

7_____ is the type of unconformity represented by the wavy line. (Figure 22.3)

8_____ million years is the minimum missing time. (Figure 22.3) Refer to Geologic Time Scale on page 84.

9_____ is the type of unconformity represented by the wavy line. (Figure 22.4)

10_____ million years is the minimum missing time. (Figure 22.4) Refer to Geologic Time Scale on page 84.

11_____ is the type of unconformity represented by the horizontal wavy line. (Figure 22.5)

12_____ million years is the minimum missing time. (Figure 22.5) Refer to Geologic Time Scale on page 84.

13_____ (100, GT 100 or LT 100) million years is the age of the shale if the intrusion is 100 million years old. (Figure 22.6)

14_____ (X, T, Q) is the oldest intrusion. (Figure 22.7)

15_____ (X, T, Q) is the youngest intrusion. (Figure 22.7)

16_____ (True or False) The intrusion occurred before the fault. (Figure 22.8)

17_____ (100, GT 100 or LT 100) million years is the age of the sandstone if the intrusion is 100 million years old. (Figure 22.8)

18_____ (100, GT 100 or LT 100) million years is the age of the fault if the intrusion is 100 million years old. (Figure 22.8)

19_____ (Rhyolite or Granite) is the older intrusion. (Figure 22.9)

20_____ (100, GT 100 or LT 100) million years is the age of the granite intrusion if the rhyolite intrusion is 100 million years old. (Figure 22.9)

21_____ is the type of unconformity. (Figure 22.10)

22_____ million years is the minimum missing time. (Figure 22.10)

23_____ million years is the minimum missing time. (Figure 22.11)

24_____ million years is the minimum missing time. (Figure 22.12)

For questions in this box five events occurred in each figure.

25_____ was the third event. (Figure 22.13)

26_____ was the fourth event. (Figure 22.14)

27_____ was the second event. (Figure 22.15)

28_____ was the fifth event. (Figure 22.16)

29_____ was the fourth event. (Figure 22.17)

30_____ was the third event. (Figure 22.18)

Refer to the Cross-section of the Grand Canyon (Figure 22.19) to answer the remaining questions:

31_____ is the type of unconformity separating the Bass Limestone and Vishnu Schist.

32_____ is the type of unconformity separating the Hakatai Shale and Tapeats Sandstone.

33_____ is the type of unconformity separating the Redwall Limestone and Temple Butte Limestone.

34_____ is the type of unconformity separating the Muav Limestone and Temple Butte Limestone.

35_____ million years is the minimum missing time between the Muav Limestone and Temple Butte Limestone.

Recalling the "Big Three Sedimentary Rocks" the two which are the most prominent cliff formers are:

36_____

37_____

and the third one which is the most prominent slope former is:

38_____

39_____ is the formation you stand on when you visit the Grand Canyon.

40_____ (1/3, 2/3, 1) mile is the best estimate for the depth of the Grand Canyon.

Age Dating 2

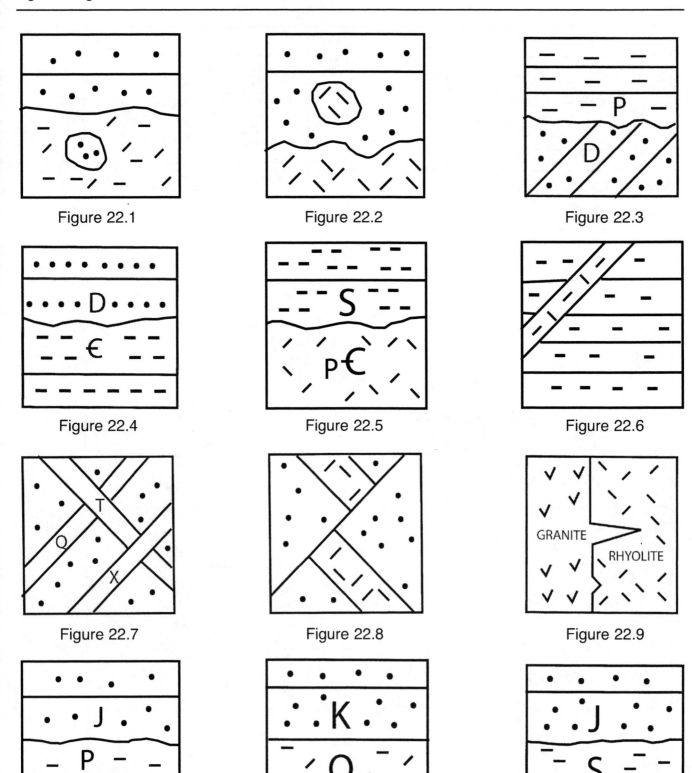

Figure 22.1

Figure 22.2

Figure 22.3

Figure 22.4

Figure 22.5

Figure 22.6

Figure 22.7

Figure 22.8

Figure 22.9

Figure 22.10

Figure 22.11

Figure 22.12

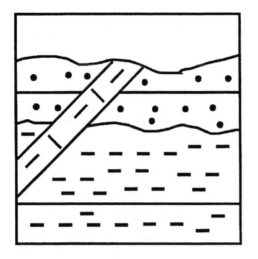

Figure 22.13

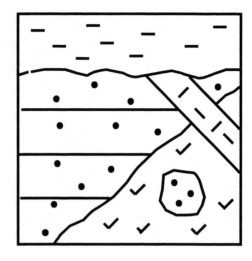

Figure 22.14

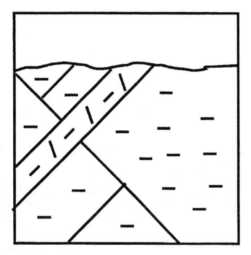

Figure 22.15

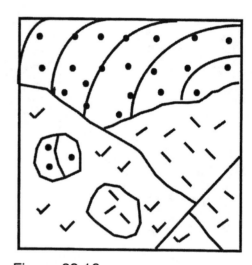

Figure 22.16

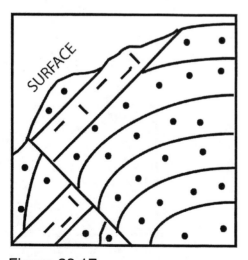

Figure 22.17

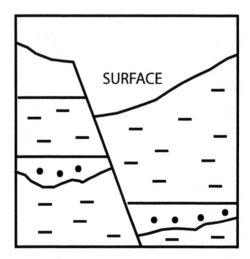

Figure 22.18

CROSS-SECTION OF THE GRAND CANYON

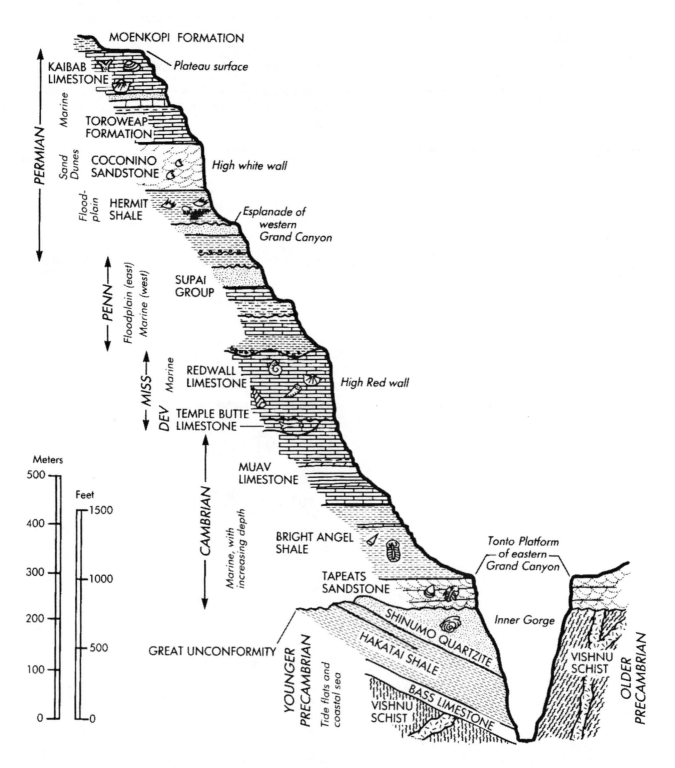

Figure 22.19 (From Chronic, *Pages of Stone*, 1st ed.)

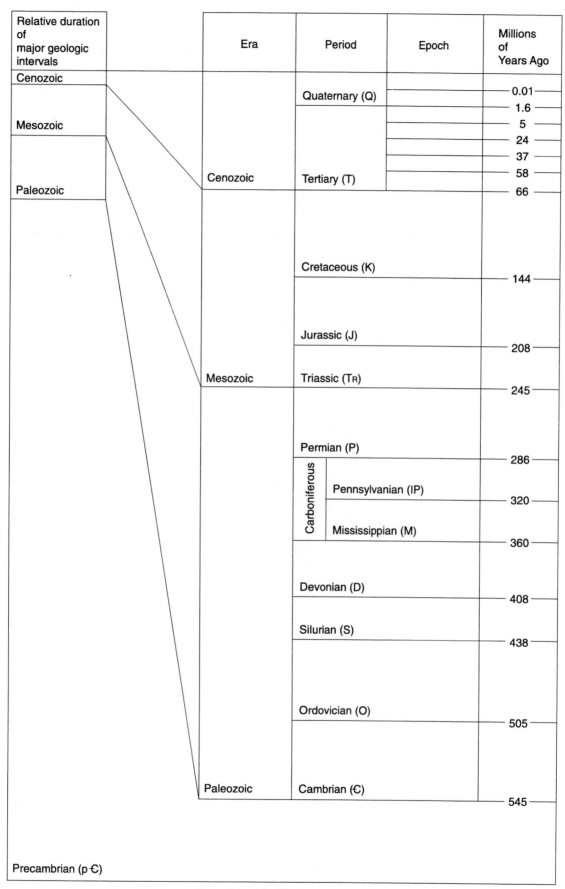

Figure 22.20 The Geologic Time Scale

SOURCE: *Laboratory Studies In Earth History,* 1st Edition, Palmer via Brice, 1983.

Exercise 23   Questions   Name:

Age Dating 3   Lab:

Complete and turn in the SCANTRON sheet. Keep these questions.
Refer to your class notes to answer the following:

1_____ atoms is the general name for the original radioactive atoms.
(A) Daughter, (B) Parent

2_____ atoms is the general name for the atoms resulting from radioactive decay. (A) Daughter, (B) Parent

3_____ atoms increase as time passes.
(A) Daughter, (B) Parent

4 _____ The more daughter atoms, the older the rock.
(A) True, (B) False

State the amount (fraction) of parent atoms remaining after the following number of half-lives have elapsed:

5_____ 1 half-life
(A) 1/2, (B) 1/4, (C) 1/8, (D) 1/16, (E) 1/32

6_____ 3 half-lives
(A) 1/2, (B) 1/4, (C) 1/8, (D) 1/16, (E) 1/32

7_____ 2 half-lives
(A) 1/2, (B) 1/4, (C) 1/8, (D) 1/16, (E) 1/32

8_____ 5 half-lives
(A) 1/2, (B) 1/4, (C) 1/8, (D) 1/16, (E) 1/32

9_____ 4 half-lives
(A) 1/2, (B) 1/4, (C) 1/8, (D) 1/16, (E) 1/32

10_____ After two half-lives none of the parent isotope remains.
(A) True, (B) False

State the number of half-lives which have elapsed when a sample has the following parent:daughter ratio:

11_____ 1:3 parent:daughter ratio
(A) 1, (B) 2, (C) 3, (D) 4, (E) 5

12_____ 1:7 parent:daughter ratio
(A) 1, (B) 2, (C) 3, (D) 4, (E) 5

13_____ 1:1 parent:daughter ratio
(A) 1, (B) 2, (C) 3, (D) 4, (E) 5

14_____ 1:15 parent:daughter ratio
(A) 1, (B) 2, (C) 3, (D) 4, (E) 5

X decays to P and has a half-life = 20 million years.

15 _____ million years is the rock age when X:P is 1:3.
(A) 20, (B) 40, (C) 60, (D) 80, (E) None

16 _____ million years is the rock age when X:P is 1:1
(A) 20, (B) 40, (C) 60, (D) 80, (E) None

17 _____ million years is the rock age when X:P is 1:7.
(A) 20, (B) 40, (C) 60, (D) 80, (E) None

Y decays to Q and has a half-life = 50 million years.

18 _____ million years is the rock age when Y:Q is 1:3.
(A) 50, (B) 100, (C) 150, (D) 200, (E) None

19 _____ million years is the rock age when Y:Q is 1:1.
(A) 50, (B) 100, (C) 150, (D) 200, (E) None

20 _____ million years is the rock age when Y:Q is 1:7.
(A) 50, (B) 100, (C) 150, (D) 200, (E) None

Z decays to R and has a half-life = 100 million years.

21 _____ million years is the rock age when Z:R is 1:7.
(A) 100, (B) 200, (C) 300, (D) 400, (E) None

22 _____ million years is the rock age when Z:R is 1:1.
(A) 100, (B) 200, (C) 300, (D) 400, (E) None

23 _____ million years is the rock age when Z:R is 1:3.
(A) 100, (B) 200, (C) 300, (D) 400, (E) None

---

24 _____ The granite is older than the dike (intrusion).
(A) True, (B) False

25 _____ million years is the age of the dike.
(A) 50, (B) GT 50, (C) LT 100, (D) 100,
(E) None

26 _____ million years is the age of the granite.
(A) 100, (B) GT 100, (C) LT 50, (D) 50,
(E) None

27 _____ is the period when the dike formed.
(Refer to page 84)
(A) Devonian, (B) Cretaceous, (C) Jurassic,
(D) Permian, (E) None

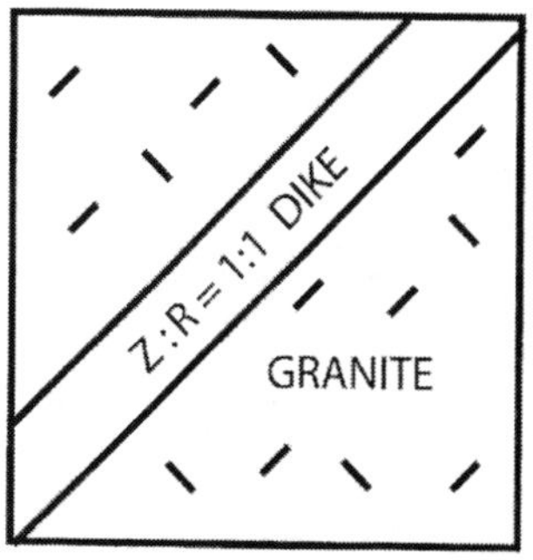

Figure 23.1

28 _____ The granite is younger than the dike (intrusion).
(A) True, (B) False

29 _____ million years is the age of the granite.
(A) 50, (B) GT 100, (C) LT 100, (D) 100,
(E) None

30 _____ million years is the age of the dike.
(A) 100, (B) GT 100, (C) LT 100, (D) 200,
(E) None

31 _____ is the period when the granite formed.
(Refer to page 84)
(A) Devonian, (B) Cretaceous, (C) Jurassic,
(D) Permian, (E) None

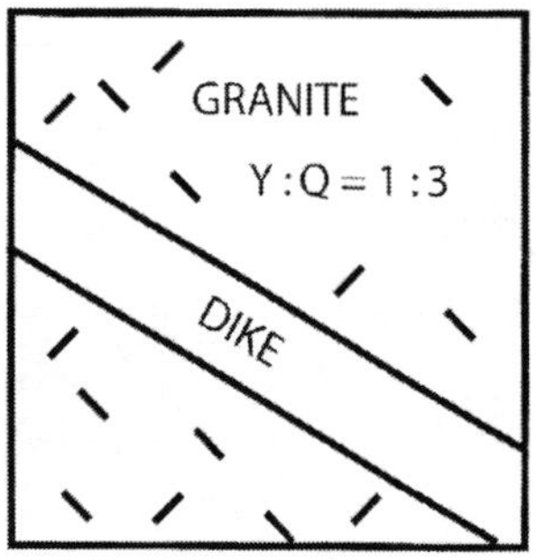

Figure 23.2

---

32 _____ million years is the age of the **shaded** dike.
(A) 20, (B) GT 20, (C) LT 20

33 _____ is the period when the **shaded** dike formed.
(A) Devonian, (B) Cretaceous, (C) Tertiary,
(D) Permian

34 _____ million years is the age of the granite.
(A) 100, (B) GT 100, (C) LT 100

35 _____ million years is the age of the unshaded dike.
(A) 20, (B) GT 100, (C) LT 20, (D) 20-100,
(E) None

36 _____ is the period when the granite formed.
(A) Devonian, (B) Cretaceous, (C) Tertiary,
(D) Permian, (E) None

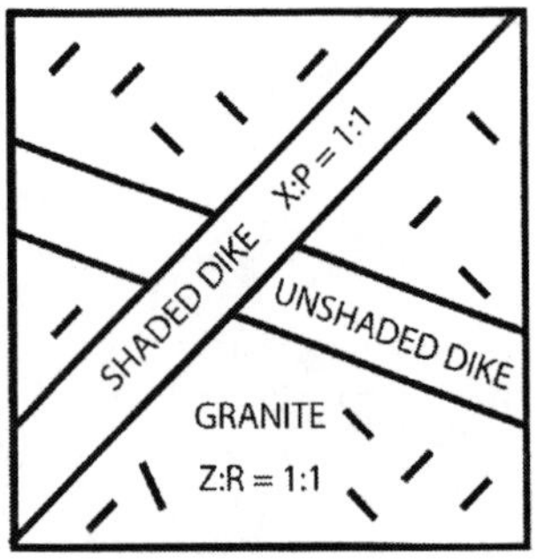

Figure 23.3

---

37 _____ million years is the age of the dike.
(A) 20, (B) LT 60, (C) GT 60, (D) 60,
(E) None

38 _____ million years is the age of the granite.
(A) 60, (B) LT 60, (C) GT 60, (D) 20,
(E) None

39 _____ is the period when the dike formed.
(A) Devonian, (B) Cretaceous, (C) Tertiary,
(D) Permian, (E) None

40 _____ million years is the age of the overlying
strata. (A) 80, (B) LT 60, (C) GT 80,
(D) 60, (E) None

41 _____ is the type of unconformity.
(A) Nonconformity, (B) Disconformity,
(C) Angular Unconformity

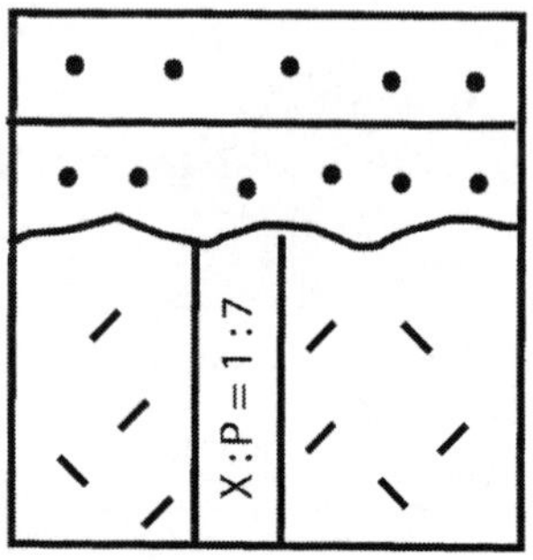

Figure 23.4

42_____ million years is the age of the lava flow.
(A) 200, (B) 400, (C) GT 400, (D) 100,
(E) None

43_____ is the period when the lava flowed.
(A) Devonian, (B) Cretaceous, (C) Tertiary,
(D) Permian, (E) None

44_____ million years is the age of the strata beneath
the lava flow. (A) 400, (B) LT 400,
(C) GT 400, (D) 200–400, (E) None

45_____ million years is the age of the strata
above the lava flow. (A) 400, (B) LT 400,
(C) GT 400

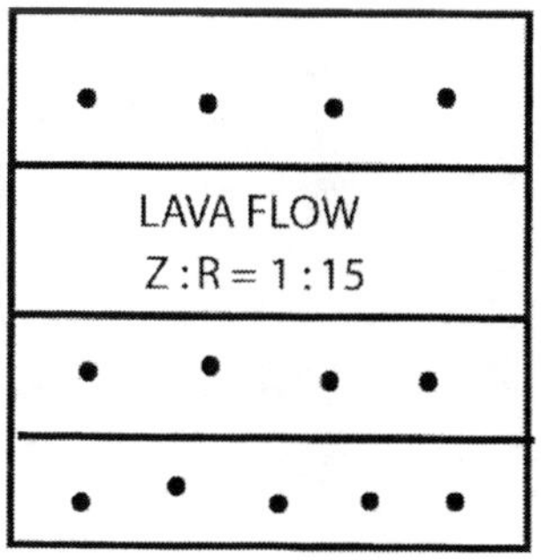

Figure 23.5

46_____ million years is the age of the granite.
(A) 200, (B) LT 200, (C) GT 200

47_____ million years is the age of the sandstone.
(A) 100, (B) LT 100, (C) GT 200, (D) 200,
(E) None

48_____ is the period when the granite formed.
(A) Devonian, (B) Cretaceous, (C) Tertiary,
(D) Jurassic, (E) None

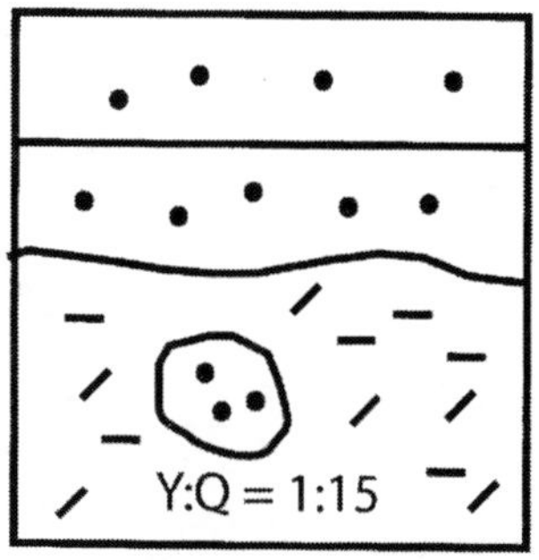

Figure 23.6

49_____ million years is the age of the granite.
(A) 200, (B) LT 400, (C) GT 500, (D) 400,
(E) None

50_____ million years is the age of the sandstone.
(A) 400, (B) LT 400, (C) GT 500,
(D) 400–500, (E) None

51_____ is the type of unconformity.
(A) Nonconformity, (B) Disconformity,
(C) Angular Unconformity

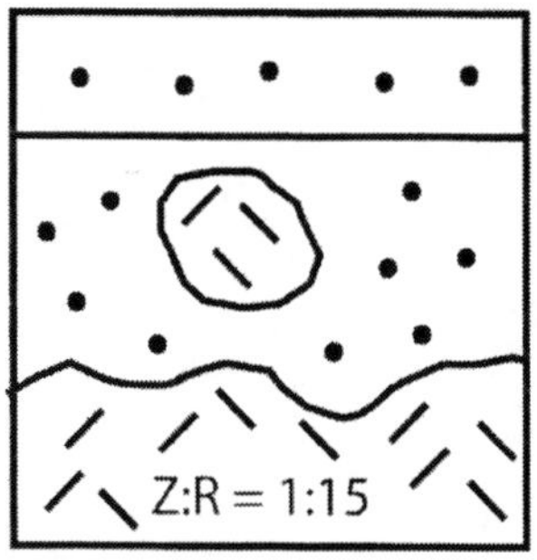

Figure 23.7

Refer to the graph on the following page (Figure 23.8):

52_____ million years is the age of the rock when the parent daughter ratio is
50:50.
(A) 60, (B) 100, (C) 160, (D) 225, (E) 330

53_____ is the geologic period when the rock with the **50:50** ratio formed.
(A) Tertiary, (B) Mississippian, (C) Jurassic, (D) Cretaceous,
(E) None

DECAY CURVE

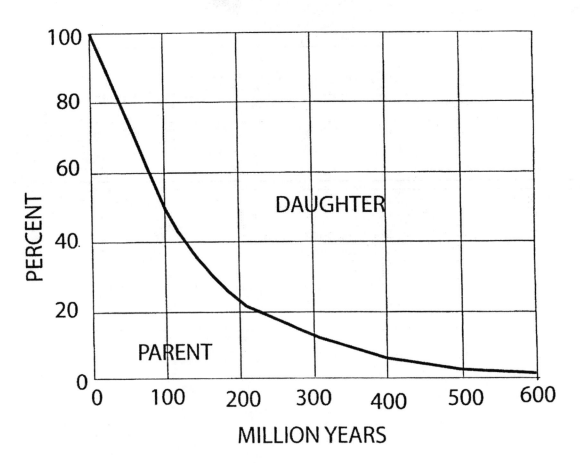

Figure 23.8

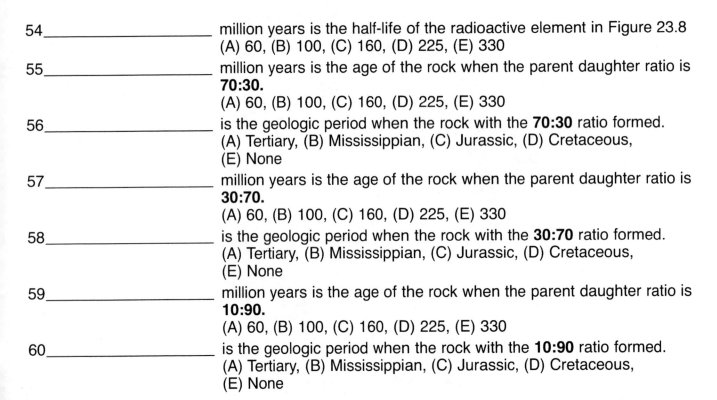

54_____ million years is the half-life of the radioactive element in Figure 23.8
(A) 60, (B) 100, (C) 160, (D) 225, (E) 330

55_____ million years is the age of the rock when the parent daughter ratio is
70:30.
(A) 60, (B) 100, (C) 160, (D) 225, (E) 330

56_____ is the geologic period when the rock with the **70:30** ratio formed.
(A) Tertiary, (B) Mississippian, (C) Jurassic, (D) Cretaceous,
(E) None

57_____ million years is the age of the rock when the parent daughter ratio is
30:70.
(A) 60, (B) 100, (C) 160, (D) 225, (E) 330

58_____ is the geologic period when the rock with the **30:70** ratio formed.
(A) Tertiary, (B) Mississippian, (C) Jurassic, (D) Cretaceous,
(E) None

59_____ million years is the age of the rock when the parent daughter ratio is
10:90.
(A) 60, (B) 100, (C) 160, (D) 225, (E) 330

60_____ is the geologic period when the rock with the **10:90** ratio formed.
(A) Tertiary, (B) Mississippian, (C) Jurassic, (D) Cretaceous,
(E) None

USING THE PERIODIC TABLE ON THE NEXT PAGE COMPLETE THE FOLLOWING:

61_____ is the atomic number of zirconium.
(A) 65, (B) 30, (C) 91, (D) 40, (E) none

62_____ is the number of neutrons of lead.
(A) 82, (B) 207, (C) 125, (D) 207.19, (E) none

63_____ is an element with similar properties as chlorine.
(A) I, (B) S, (C) Ar, (D) O, (E) none

64_____ is the atomic number of mercury.
(A) 201, (B) 200, (C) 121, (D) 80, (E) none

65_____ is the number of neutrons of zirconium.
(A) 40, (B) 91, (C) 51, (D) 35, (E) none

66_____ is an element with similar properties as phosphorus.
(A) N, (B) Si, (C) S, (D) Po, (E) none

67_____ is the atomic mass of bromine.
(A) 209, (B) 83, (C) 35, (D) 80, (E) none

68_____ is the atomic number of magnesium.
(A) 27, (B) 59, (C) 29, (D) 64, (E) none

69_____ is the number of neutrons of radium.
(A) 88, (B) 226, (C) 136, (D) 138, (E) none

70_____ is an element with similar properties as calcium.
(A) Ba, (B) Sc, (C) k, (D) Li, (E) none

Modern Periodic Table

MODERN PERIODIC TABLE

The number of electrons in filled shells is shown in the column at the extreme left; the remaining electrons for each element are shown below the symbol and atomic number for each element. The atomic weights shown above the symbols are based on Carbon-12.

METALS

NONMETALS

TRANSITION METALS

PERIODS	I A	II A	III B	IV B	V B	VI B	VII B	VIII	VIII	VIII	I B	II B	III A	IV A	V A	VI A	VII A	0
1 (0)	1.00797 H [1] 1																	4.0026 He [2] 2
2 (2)	6.939 Li [3] 1	9.0122 Be [4] 2											10.811 B [5] 3	12.01115 C [6] 4	14.0067 N [7] 5	15.9994 O [8] 6	18.9984 F [9] 7	20.183 Ne [10] 8
3 (2,8)	22.9898 Na [11] 1	24.312 Mg [12] 2											26.9815 Al [13] 3	28.086 Si [14] 4	30.9738 P [15] 5	32.064 S [16] 6	35.453 Cl [17] 7	39.948 Ar [18] 8
4 (2,8)	39.102 K [19] 8,1	40.08 Ca [20] 8,2	44.956 Sc [21] 9,2	47.90 Ti [22] 10,2	50.942 V [23] 11,2	51.996 Cr [24] 13,1	54.9380 Mn [25] 13,2	55.847 Fe [26] 14,2	58.9332 Co [27] 15,2	58.71 Ni [28] 16,2	63.54 Cu [29] 18,1	65.37 Zn [30] 18,2	69.72 Ga [31] 18,3	72.59 Ge [32] 18,4	74.9216 As [33] 18,5	78.96 Se [34] 18,6	79.909 Br [35] 18,7	83.80 Kr [36] 18,8
5 (2,8,18)	85.47 Rb [37] 8,1	87.62 Sr [38] 8,2	88.905 Y [39] 9,2	91.22 Zr [40] 10,2	92.906 Nb [41] 12,1	95.94 Mo [42] 13,1	(99) Tc [43] 14,1	101.07 Ru [44] 15,1	102.905 Rh [45] 16,1	105.4 Pd [46] 18	107.870 Ag [47] 18,1	112.40 Cd [48] 18,2	114.82 In [49] 18,3	118.69 Sn [50] 18,4	121.75 Sb [51] 18,5	127.60 Te [52] 18,6	126.9044 I [53] 18,7	131.30 Xe [54] 18,8
6 (2,8,18)	132.905 Cs [55] 18,8,1	137.34 Ba [56] 18,8,2	[57-71] *	178.49 Hf [72] 32,10,2	180.948 Ta [73] 32,11,2	183.85 W [74] 32,12,2	186.2 Re [75] 32,13,2	190.2 Os [76] 32,14,2	192.2 Ir [77] 32,15,2	195.09 Pt [78] 32,17,1	196.967 Au [79] 32,18,1	200.59 Hg [80] 32,18,2	204.37 Tl [81] 32,18,3	207.19 Pb [82] 32,18,4	208.980 Bi [83] 32,18,5	(210) Po [84] 32,18,6	(210) At [85] 32,18,7	(222) Rn [86] 32,18,8
7 (2,8,18,32)	(223) Fr [87] 18,8,1	(226.05) Ra [88] 18,8,2	[89-103] †	[104]	[105]	[106]	[107]	[108]										

* LANTHANIDE SERIES

La [57]	Ce [58]	Pr [59]	Nd [60]	Pm [61]	Sm [62]	Eu [63]	Gd [64]	Tb [65]	Dy [66]	Ho [67]	Er [68]	Tm [69]	Yb [70]	Lu [71]
138.91 18,9,2	140.12 20,8,2	140.907 21,8,2	144.24 22,8,2	(145) 23,8,2	150.35 24,8,2	151.96 25,8,2	157.25 25,9,2	158.924 27,8,2	162.50 28,8,2	164.930 29,8,2	167.26 30,8,2	168.934 31,8,2	173.04 32,8,2	174.97 32,9,2

† ACTINIDE SERIES

Ac [89]	Th [90]	Pa [91]	U [92]	Np [93]	Pu [94]	Am [95]	Cm [96]	Bk [97]	Cf [98]	Es [99]	Fm [100]	Md [101]	No [102]	Lw [103]
(227) 18,9,2	232.038 18,10,2	(231) 20,9,2	238.03 21,9,2	(237) 23,8,2	(242) 24,8,2	(243) 25,8,2	(245) 25,9,2	(245) 26,9,2	(248) 28,8,2	(253) 29,8,2	(254) 30,8,2	(256) 31,8,2	(253) 32,8,2	(257) 32,9,2

SOURCE: Nebergall, et al., *General Chemistry.* Copyright © 1963 by D.C. Heath & Company.

Microfossil Preparation

Note: All instructions below are "about" because it does not matter that much. This is not chemistry lab—you do not need to be precise.

1. **Only file on the lab carts.** You may use any of the rocks on the lab carts or one you brought. File your rock onto a sheet of paper until you have about 1/4 of a little plastic beaker full.

2. Take a big steel beaker. The beaker can be muddy—you do not need to clean it. **Careful: Gently** turn on the water (It comes out like a fire hose!!!) Fill it about 1/3 full of water and dump your Austin Chalk powder into it. Put it in a ring on a hotplate and set the dial on "9". Drop a couple of pieces of gravel size chalk into the beaker to let it boil more gently. Bring the water to a boil and boil it about 20 minutes to help break up the chalk. Share a hotplate with someone—i.e., two beakers per hotplate. Place the hotplate far away from your face.

3. While the water is boiling you can learn how to use the microscope (look at your fingers, ring, etc.) **Use both hands to carry the microscope.** Make a special trip just for the microscope.

4. **Do not let the microscope cord or hotplate cord dangle below the table top** where you can get your legs and lab stools caught in them unless you want to buy the microscope or pour boiling water in your lap. You can put something heavy on the cords to keep them from dangling. This is your chance to use one of those huge books they made you buy for another course like accounting or British Lit.

5. After the water has boiled about 20 minutes: (1) **turn-off** the hot plate and (2) take a pair of beaker tongs, firmly clamp them around the beaker near the top. Take it to any of the four sinks with hoses on the faucets.

6. If the coarse sieve has gravel in it dump the gravel into the trash can (**not** in the sink!!!). Leave the sieves setting in the sink. **Be sure the fine sieve is on the bottom!** Pour the beaker water into the sieves as shown:

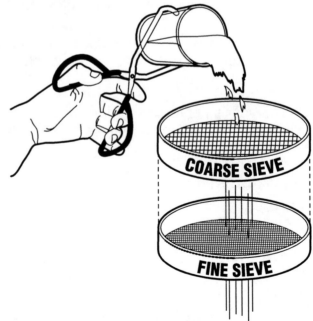

COARSE SIEVE

FINE SIEVE

Figure 24.1

7. Rinse out the beaker a few times into the sieves. Set the beaker aside, take the rubber hose and gently wash the sediment in the top sieve.

8. Set the top sieve aside. Take the fine sieve (bottom sieve hopefully) and gently and **thoroughly** wash the sand.

9. Ask your lab partner to check to see if the beaker is cool enough to handle. If yes then wash the beaker until it is perfectly **clean.**

10. Now take the bottom sieve and wash the sand over to one side. Set the clean beaker in the sink and then flush out the sediment into the beaker from the backside as shown here:

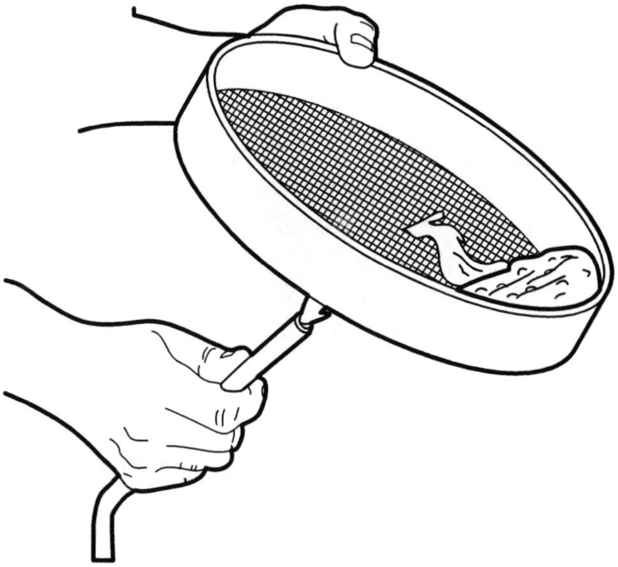

11. If the water in the beaker looks murky, pour off some, add fresh water, pour off some, etc. until it looks clear. Then pour off the excess water leaving only sand.

12. Place the beaker on a **cool hotplate** and turn the setting to about "5". If the sand starts popping then the hotplate is too hot and you must take the beaker off the hotplate and turn down the setting.

Figure 24.2

13. **Do not stir the sand.** It will only stick it together and make a mess. When the sand is dry it will be lighter in color and by swirling the beaker the sand will be loosened. **Do not make noise bumping the beaker on the table— use a book;** here again, British literature and accounting come to mind.

14. Pour the dry sand into your plastic pill bottle.

15. Sprinkle the sand **one layer thick** on an aluminum tray. **Note:** The casual observer (your lab partner regrettably) will dump the whole bottle into the aluminum tray to form a sand dune!

16. Examine your sample with the microscope. Next show it to your instructor but **expect the words: "You need to make another sample!"** That is part of the mystique of micropaleontology. You are expected to prepare several samples to complete this microfossil slide.

Microfossil Collecting

1. Generally speaking you need one of everything on the supply cart.

2. Take the little cardboard microfossil slide and coat row one (boxes 1–12) with glue several times. The glue is in the tiny plastic bottles.

3. **To see what the forams look like refer to pages 97 and 98.**

4. Set you microscope on the higher magnification. Sprinkle your sand onto the little black metal tray **only one layer thick** and examine it by moving the tray, not by raking through it. When you see a foram then isolate it by pushing the surrounding sand away with the needle. Do not alter the tip of the needle. Get another if necessary. To locate the needle while looking through the microscope wave it until you see it flash by. Then back it up and lower it to the foram:

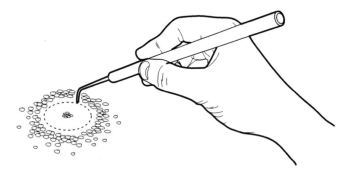

Figure 25.1

5. Wash out the little plastic beaker, put water in it and wet the brush. Gently wipe off the excess water between two fingers. To locate the brush while looking through the microscope wave it until you see it flash by. Then back it up and lower it to the foram and they will stick together!

6. Remove the aluminum tray and place the microslide in its place. While looking through the microscope lower the foram to the correct box **(refer to page 99 for correct location to place the forams on the slide).** If you are not sure then make an educated guess; you can always move it later because the glue is water soluble. Leave the foram just barely surrounded with water:

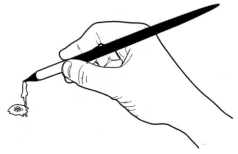

Figure 25.2

The water will dissolve the glue and when the glue dries again the foram should stick. If in doubt then put a little glue on the fossil. If you decide to move the foram later just reverse the process: Add a little water then wait a few seconds until the glue dissoves. Store your microslide in a pill bottle or something safe.

7. At the end of lab please return everything to its place of origin and clean your area. **Make a special trip just for the microscope.**

NOTE: You will be graded on the quality of your specimens. You should only collect complete specimens, free of rock.

REQUIRED FORAMS:

Genus: *Planulina*

(PLAN-U-LIE-NAH)

YOU WILL NEED 10:
5 FOR BOX 1
5 FOR BOX 4

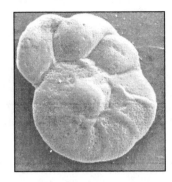

Figure 25.3

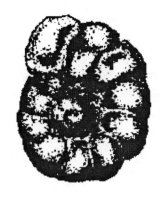

Figure 25.4

Genus: *Globigerina*

(GLOW-BEH-GEE-RI-NAH)

YOU WILL NEED 10:
5 FOR BOX 2
5 FOR BOX 5

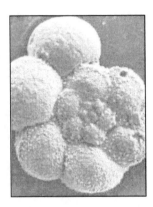

Figure 25.5

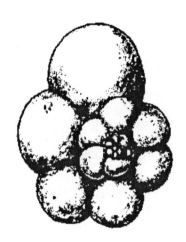

Figure 25.6

Genus: *Heterohelix*

(HET-ER-OH-HE-LIX)

YOU WILL NEED 10:
5 FOR BOX 3
5 FOR BOX 6

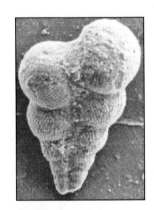

Figure 25.7

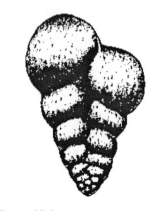

Figure 25.8

SOURCE: Figures 25.4, 25.6, and 25.8 from *The Handbook of Cretaceous Foraminifera of Texas* by Don L. Frizzell. Copyright © 1954 by the Bureau of Economic Geology. Reprinted by permission.
SOURCE: Figures 25.3, 25.5, and 25.7 SEM photography by Dr. Ignacio Pujana.

FORAMS OF ANOTHER GENUS:

YOU WILL ONLY NEED:
1 FOR BOX 7
1 FOR BOX 8

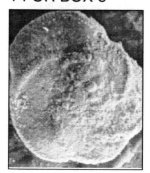

Figure 25.9
Genus: *Globotruncana*

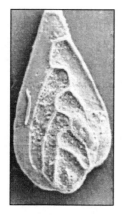

Figure 25.10
Genus: *Kyphopyxa*

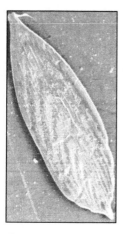

Figure 25.11
Genus: *Frondicularia*

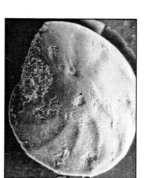

Figure 25.12
Genus: *Robulus*

FOR OTHER
POSSIBILITIES REFER
TO THE:

HANDBOOK OF
CRETACEOUS
FORMAMINIFERA OF
TEXAS

FOR OTHER
POSSIBILITIES REFER
TO THE:

HANDBOOK OF
CRETACEOUS
FORMAMINIFERA OF
TEXAS

OSTRACODS:

YOU WILL RECALL THAT OSTRACODS ARE ARTHROPODS, NOT FORAMS.

YOU WILL ONLY NEED:
1 FOR BOX 10
1 POSSIBLY FOR BOXES 7 OR 8 AS SUBSTITUES OF FORAMS OF ANOTHER GENUS

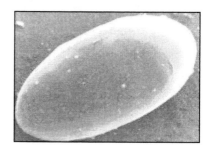

Figure 25.13
Ostracod

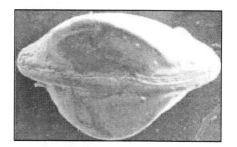

Figure 25.14
Ostracod

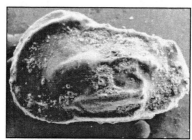

Figure 25.15
Ostracod

SOURCE: Figures 25.9 to 25.15 SEM photography by Dr. Ignacio Pujana.

Report Sheet   Name:

Microfossil Collecting   Lab:

Turn in this sheet with your finished microfossil slide.
─────────────── **Complete Boxes 1, 2 & 3:** ───────────────

5 <u>Planulina</u> (@5 points each = 25 points)

5 <u>Globigerina</u> (@5 points each = 25 points)

5 <u>Heterohelix</u> (@5 points each = 25 points)

─────────── **Complete <u>any 5 of the 7</u> boxes below:** ───────────

5 additional <u>Planulina</u> (@1 point each = 5 points)

5 additional <u>Globigerina</u> (@1 point each = 5 points)

5 additional <u>Heterohelix</u> (@1 point each = 5 points)

1 foram of another genus or ostracod (5 points)

1 foram of another genus or ostracod (5 points)

5 forams of the same genus of your choice
                    (@1 point each = 5 points)

1 ostracod (5 points)

APPENDIX

Maps

Determining Coordinates

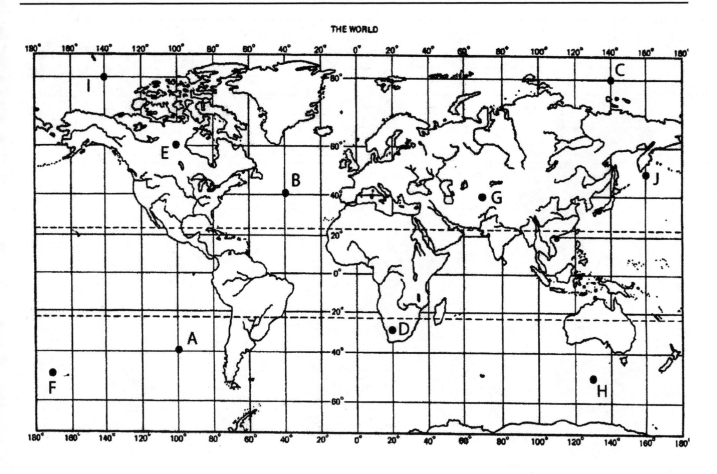

THE WORLD

Determine the latitude and longitude for each location:

A   40 deg S   ,   100 deg W   .   F   50 deg S   ,   170 deg W   .

B   40 deg N   ,   40 deg W   .   G   40 deg N   ,   70 deg E   .

C   80 deg N   ,   140 deg E   .   H   50 deg S   ,   130 deg E   .

D   30 deg S   ,   20 deg E   .   I   80 deg N   ,   140 deg W   .

E   60 deg N   ,   100 deg W   .   J   50 deg N   ,   160 deg E   .

(THE ANSWERS ARE GIVEN HERE SO YOU CAN PRACTICE.)

Map SOURCE: Adapted from *Physical Geography and Earth Science* by Ray and James, 1970.

The Geographic Regions of Texas

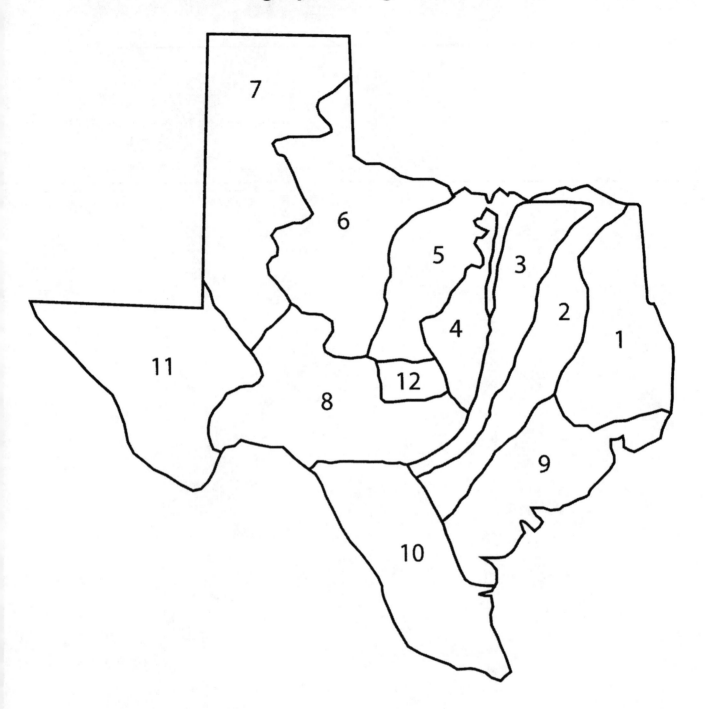

Mineral Facts List

APATITE . . . .(5) . . . . H = 5

AUGITE . . . .(24) . . . . FERROMAGNESIAN
(AW-JITE)   TWO CLEAVAGES AT RIGHT ANGLES

AZURITE. . . . . AZURE BLUE (INHERENT COLORATION)
                                  COPPER ORE
                                  SECONDARY MINERAL

BARITE . . . .(30) . . . . BLADED CRYSTALS FORM ROSE PATTERNS
                                  HIGH SPECIFIC GRAVITY

BIOTITE . . . .(22) . . . . BLACK MICA
                                  ONE CLEAVAGE
                                  FERROMAGNESIAN

CALCITE . . . .(3) . . . . RHOMBOHEDRAL CLEAVAGE   BEFORE   AFTER
                                  LIMESTONE IS MADE OF CALCITE
                                  MARBLE IS MADE OF CALCITE
                                  H = 3
                                  DISPLAYS DOUBLE REFRACTION
                                  EFFERVESCES IN HYDROCHLORIC ACID
                                  $CaCO_3$ (CALCIUM CARBONATE)   Figure 5.1 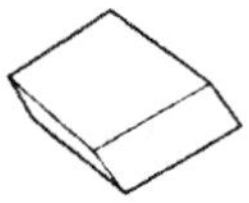   Figure 5.2

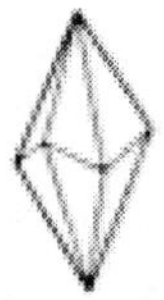

CHALCOPYRITE . .(27) . . . . FOOL'S GOLD
(CAL-CO-PI-RITE)   COPPER ORE
                                  $CuFeS_2$

CORUNDUM . . . . . .(9) . . . . IF GEM QUALITY AND RED IS RUBY
                                  IF GEM QUALITY AND BLUE IS SAPPHIRE
                                  H = 9
                                  HIGH SPECIFIC GRAVITY
                                  ABRASIVE DUE TO HIGH HARDNESS
                                  Al_2O_3

DIAMOND . . . .(10) . . . . C (ENTIRELY CARBON) _____   BEFORE   AFTER
                                  OCTAHEDRAL CRYSTALS
                                  OCTAHEDRAL CLEAVAGE
                                  H = 10
                                  ABRASIVE DUE TO HIGH HARDNESS
                                                                      Figure 5.3 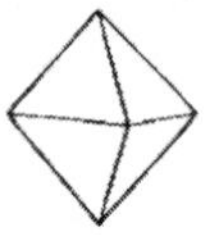   Figure 5.4

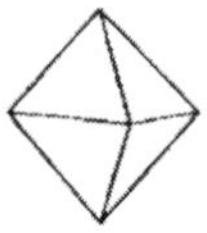

FLUORITE . . . .(4) . . . . USUALLY PURPLE BUT EXOTIC COLORATION
                                  H = 4 _____   BEFORE   AFTER
                                  CUBIC CRYSTALS
                                  OCTAHEDRAL CLEAVAGE
                                  FORMS "TRIANGLES" WHEN IT BREAKS
                                  CaF_2
                                                                      Figure 5.5 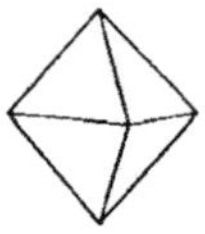   Figure 5.6

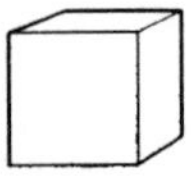

GALENA . . . .(28) . . . . LEAD ORE
CUBIC CRYSTALS
CUBIC CLEAVAGE
HIGH SPECIFIC GRAVITY
PbS

BEFORE   AFTER

Figure 5.7   Figure 5.8

GARNET . . . .(31) . . . . DODECAHEDRAL CRYSTALS
ABRASIVE

BEFORE

Figure 5.9

GRAPHITE . . . . C (ENTIRELY CARBON)
SLIPPERY

GYPSUM . . . .(2) . . . . H = 2
PLASTER OF PARIS
USED TO MAKE WALLBOARD
BLADED CRYSTALS FORM ROSE PATTERNS
SELENITE IS THE VARIETY IN THE TRAY

HALITE . . . .(32) . . . . TABLE SALT
CUBIC CRYSTALS
CUBIC CLEAVAGE
NaCl (SODIUM CHLORIDE)

BEFORE   AFTER

Figure 5.10   Figure 5.11

HEMATITE . . . . . .(33) . . . . RED STREAK
IRON ORE
Fe_2O_3 (IRON OXIDE)

HORNBLENDE . . .(23) . . . . FERROMAGNESIAN
TWO CLEAVAGES FORMING DIAMOND END
ELONGATE

KAOLINITE . . . . . .(20) . . . . CLAY
FORMED BY WEATHERING FELDSPAR
USED TO MAKE BATHROOM FIXTURES
SECONDARY MINERAL (FORMS BY WEATHERING OTHER
MINERALS)

LIMONITE . . . . RUST (RUST ITSELF IS LIMONITE)
SECONDARY MINERAL

MAGNETITE . . . . .(34) . . . . IRON ORE
MAGNETIC (ATTRACTS A MAGNET)
HIGH SPECIFIC GRAVITY
Fe_3O_4

MALACHITE . . . . COPPER ORE
BRIGHT GREEN (INHERENT COLORATION)
SECONDARY MINERAL

MUSCOVITE ....(21) .... WHITE MICA
ONE PERFECT CLEAVAGE
COLORLESS IF THIN
WAS USED FOR WINDOW GLASS

OLIVINE ....(25) .... FERROMAGNESIAN
GEM QUALITY IS CALLED PERIDOT

ORTHOCLASE ..(16–17)... FELDSPAR
USUALLY PINK
MAIN MINERAL IN GRANITE
H = 6
TWO CLEAVAGES AT RIGHT ANGLES
"ORTHO" MEANS RIGHT, "CLASE" MEANS BREAKS
"ORTHOCLASE" MEANS BREAKS AT RIGHT ANGLES
$KAlSi_3O_8$

PLAGIOCLASE ..(18–19)... FELDSPAR
TWO CLEAVAGES AT RIGHT ANGLES
USUALLY WHITE
SOMETIMES IRIDESCENT PURPLE
H = 6
SOMETIMES DISPLAYS STRIATIONS ON CLEAVAGE PLANES

PYRITE ....(26) .... CRYSTALS ARE CUBES OR PYRITOHEDRONS
FOOL'S GOLD
WHITISH GOLD COLOR
FeS_2

BEFORE & BEFORE

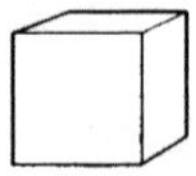

Figure 5.12 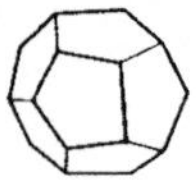   Figure 5.13

QUARTZ ....(7) .... SIX-SIDED CRYSTALS
H = 7
ABRASIVE (EX: SANDPAPER)
EXOTIC COLORATION:
PURPLE = AMETHYST
PINK = ROSE QUARTZ
WHITE = MILKY QUARTZ
BLACK = SMOKY QUARTZ
NO CLEAVAGE
SiO_2

BEFORE

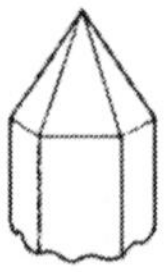

Figure 5.14

SPHALERITE ....(29) .... ZINC ORE
YELLOW STREAK WHICH IS SMELLY
ZnS

TALC ....(1) .... H = 1
SLIPPERY

TOPAZ ....(8) .... TEXAS STATE GEM
HIGH SPECIFIC GRAVITY
ABRASIVE, H = 8

(All Mineral Illustrations from Ossian, *Insights in Earth Science*, 2nd ed.)